# Cosmotechnics



This volume is initial reflections on the meaning and the implications of Yuk Hui's notion of cosmotechnics, which opens up an anti-universalist and pluralist perspective on technology beyond the West.

Martin Heidegger's famous analysis of the essence of technology as *enframing* and as rooted in ancient Greek *techne* has had a crucial influence on the understanding and critique of technological society and culture in the twentieth century. However, it is still unclear to what extent his analysis can also be applied to the development of technology outside of 'the West', e.g. in China, Africa, and Latin America, particularly against the backdrop of receding Western domination and impending global ecological disaster. Acknowledging the planetary expansion of Western technology already observed by Heidegger, yet also recognizing the existence of non-Western origins of technical relationships to the cosmos, Yuk Hui's notion of cosmotechnics calls for a rethinking – in dialogue with decolonial studies and the so-called ontological turn in contemporary anthropology – of the question concerning technology which challenges the universality still present in Heidegger (as well as in Simondon and Stiegler) and proposes a radical technological or rather cosmotechnical pluralism or technodiversity. The contributors to this volume critically engage with this proposal and examine the possible implications of Hui's cosmotechnical turn in thinking about technology as it becomes a planetary force in our current age of the Anthropocene.

The chapters in this book were originally published as a special issue of *Angelaki*.

**Yuk Hui** teaches at the City University of Hong Kong. He did his Ph.D. thesis at Goldsmiths College in London, postdoctoral studies in France, and Habilitation thesis in Germany. Between 2012 and 2019 he taught at Leuphana University and Bauhaus University in Germany. Hui's research focus is on the philosophy of technology.

**Pieter Lemmens** teaches philosophy and ethics at Radboud University, Nijmegen, the Netherlands. His current interests are the philosophical and politico-economic aspects of human (cognitive) enhancement technologies, the philosophical aspects of trans- and post-humanism, philosophy of psychedelics, and philosophy of technology in the age of the Anthropocene.

# Cosmotechnics

## For a Renewed Concept of Technology in the Anthropocene

***Edited by***
**Yuk Hui and Pieter Lemmens**

Routledge
Taylor & Francis Group
LONDON AND NEW YORK

First published 2021
by Routledge
2 Park Square, Milton Park, Abingdon, Oxon, OX14 4RN

and by Routledge
605 Third Avenue, New York, NY 10158

*Routledge is an imprint of the Taylor & Francis Group, an informa business*



*British Library Cataloguing-in-Publication Data*
A catalogue record for this book is available from the British Library

ISBN13: 978-0-367-76936-9 (hbk)
ISBN13: 978-0-367-76937-6 (pbk)
ISBN13: 978-1-003-16900-0 (ebk)

Typeset in Minion Pro
by codeMantra

**Publisher's Note**
The publisher accepts responsibility for any inconsistencies that may have arisen during the conversion of this book from journal articles to book chapters, namely the inclusion of journal terminology.

# Contents

# Citation Information

The chapters in this book were originally published in *Angelaki*, volume 25, issue 4 (August 2020). When citing this material, please use the original page numbering for each article, as follows:

**Preface**
*Cosmotechnics*
Yuk Hui
*Angelaki*, volume 25, issue 4 (August 2020) pp. 1–2

**Introduction**
*Cosmotechnics and the ontological turn in the age of the Anthropocene*
Pieter Lemmens
*Angelaki*, volume 25, issue 4 (August 2020) pp. 3–8

**Chapter 1**
*Other turnings. Yuk Hui's pluralist cosmotechnics in between heidegger's ontological and stiegler's organological understanding of technology*
Pieter Lemmens
*Angelaki*, volume 25, issue 4 (August 2020) pp. 9–25

**Chapter 2**
*Cosmotechnics from an anthropotechnological perspective*
Marco Pavanini
*Angelaki*, volume 25, issue 4 (August 2020) pp. 26–38

**Chapter 3**
*Neosubstantivism as cosmotechnics: Gilbert Simondon versus the transhumanist synthesis*
Andrés Vaccari
*Angelaki*, volume 25, issue 4 (August 2020) pp. 39–53

**Chapter 4**
*Machine and ecology*
Yuk Hui
*Angelaki*, volume 25, issue 4 (August 2020) pp. 54–66

**Chapter 5**

*Noodiversity, technodiversity. Elements of a new economic foundation based on a new foundation for theoretical computer science*
Bernard Stiegler & translated by Daniel Ross
*Angelaki*, volume 25, issue 4 (August 2020) pp. 67–80

**Chapter 6**

*Technics and agency. The pluralism and diversity of* technē
Jason Tuckwell
*Angelaki*, volume 25, issue 4 (August 2020) pp. 81–96

**Chapter 7**

*Philosophy in the light of AI. Hegel or Leibniz*
Sjoerd van Tuinen
*Angelaki*, volume 25, issue 4 (August 2020) pp. 97–109

**Chapter 8**

*Towards a fifth ontology for the Anthropocene*
Clive Hamilton
*Angelaki*, volume 25, issue 4 (August 2020) pp. 110–119

**Chapter 9**

*The black angel of history. Afrofuturism's cosmic techniques*
Frédéric Neyrat & translated by Daniel Ross
*Angelaki*, volume 25, issue 4 (August 2020) pp. 120–134

**Chapter 10**

*Equivocations of the body and cosmic arts. An experiment in polyrealism*
Peter Skafish
*Angelaki*, volume 25, issue 4 (August 2020) pp. 135–148

# Contributors

**Clive Hamilton** is Professor of Public Ethics at Charles Sturt University, Canberra, Australia, and has held various visiting academic positions, including at the University of Oxford, UK; Sciences Po, France; the University of Heidelberg, Germany; and Yale University, USA. Prior to his appointment at Charles Sturt University, he was for several years the Executive Director of a progressive think tank, the Australia Institute. His books include the trilogy *Requiem for a Species: Why we Resist the Truth about Climate Change* (2010), *Earthmasters: Playing God with the Climate* (2013), and *Defiant Earth: The Fate of Humans in the Anthropocene* (2017).

**Yuk Hui** wrote his Ph.D. thesis under the French philosopher Bernard Stiegler (1952–2020) at Goldsmiths College in London, and received his Habilitation from Leuphana University in Germany. He is an Associate Professor at the City University of Hong Kong, and Visiting Professor of Philosophy and Technology at the China Academy of Art. His monographs include *On the Existence of Digital Objects* (2016), *The Question Concerning Technology in China. An Essay in Cosmotechnics* (2016, 2019), *Recursivity and Contingency* (2019), and forthcoming *Art and Cosmotechnics* (2021). His books have been endorsed by The Philosophical *Quarterly, Radical Philosophy, Jahrbuch Technikphilosophie, Frankfurter Allgemeine Zeitung, Folha de Sao Paulo* among others, and have been translated into French, German, Italian, Spanish, Portuguese, Chinese, Japanese, Korean and Russian.

**Pieter Lemmens** teaches philosophy and ethics at Radboud University, Nijmegen, the Netherlands. He has published on themes in the philosophy of technology and new media; on the work of Martin Heidegger, Peter Sloterdijk and Bernard Stiegler; on the Anthropocene as well as on post-operaist Marxism (Hardt, Negri, Berardi) and topics related to philosophical anthropology and (post)phenomenology. He translated Stiegler's *Philosopher par accident in Dutch* (2014) and co-edited a book on the philosophy of landscape and place (2011) as well as a volume on contemporary German philosophy (2013), both in Dutch. He is also the editor of the forthcoming *Entangled. An Introduction to the Philosophy of Technology* (Boom, 2021).

**Frédéric Neyrat** is French philosopher, Associate Professor, and Mellon-Morgridge Professor of Planetary Humanities in the English department of UWMadison (USA). He is the editor of the electronic platform *Alienocene* and a member of the editorial boards of the journals *Lignes*, *Multitudes*, and *Terrestres*. Recently, he published *Atopias: Manifesto for a Radical Existentialism* (Fordham, 2017), *The Unconstructable Earth: An Ecology of Separation* (Fordham, 2018), and *Literature and Materialisms* (Routledge, 2020).

**Marco Pavanini** is Ph.D. student at Durham University, UK, alongside the Centre for Culture and Ecology. His research interests concern the relationship between technics and human existence, with a particular focus, on the one hand, to the role played by technics within anthropogenesis; on the other, to the effect contemporary technologies have on the processes of formation of collectives and of a sense of belonging. More broadly, he is interested in contemporary philosophy, with special reference to its relationship with the sciences and the thought of Bernard Stiegler and Peter Sloterdijk.

**Daniel Ross** obtained his doctorate from Monash University, Melbourne, Australia, in 2002 with a thesis on Heidegger and the political. His first book, *Violent Democracy* (Cambridge UP), was published in 2004, and the same year saw the premier at the Rotterdam International Film Festival of his award-winning feature documentary entitled *The Ister*. He has worked extensively with Stiegler and published eleven volumes of translation of Stiegler's work, most recently *The Age of Disruption: Technology and Madness in Computational Capitalism* (Polity, 2019) and *Nanjing Lectures 2016–2019* (Open Humanities, 2020). His latest book, *Political Anaphylaxis*, is published by Open Humanities Press (2021).

**Peter Skafish** has held research and teaching positions at the Collège de France, at the University of California, Berkeley, USA; McGill University, Montreal, Canada; and the Bauhaus-Universität, Weimar, Germany; and is currently developing The Institute of Speculative and Critical Inquiry, a research centre intended to help invent alternative forums for the humanities as the US university system recedes. He is the translator of Eduardo Viveiros de Castro's *Cannibal Metaphysics*, Catherine Malabou's *The Heidegger Change*, and the co-editor of *Comparative Metaphysics*. He wrote, most recently, the introduction to Claude Lévi-Strauss's *From Montaigne to Montaigne* and is currently completing a book called *Rough Metaphysics*.

**Bernard Stiegler** (1952–2020) was the President of the Institut de Recherche et d'Innovation at the Centre Pompidou, Paris, France, and Founder of the association Ars Industrialis. After obtaining his Ph.D. at the École des haute études en sciences sociales under the supervision of Jacques Derrida, he has published over thirty books and taught at numerous universities, including many years at the Université de technologie de Compiègne, France. His most recent book is *Qu'appelle-t-on Panser? 2. La leçon de Greta Thunberg* (Les Liens qui Libèrent, 2020), and his most recent work translated into English is *Nanjing Lectures 2016–2019* (Open Humanities, 2020).

**Jason Tuckwell** teaches literature and theory at Western Sydney University, Australia. His research focuses on aesthetics and technology and particularly upon problems of creative and technical praxes. His recent publications include articles in *Philosophy Today* (2020, 2019) and *Transcultural Studies* (2018), and a monograph *Creation and the Function of Art: Technē, Poiesis and the Problem of Aesthetics* (Bloomsbury Studies in Continental Philosophy, 2017).

**Andrés Vaccari** is Associate Researcher at the National University of Río Negro and the National Scientific and Technical Research Council. He has a Ph.D. in Philosophy from Macquarie University, Sydney, Australia, and has taught at the universities of Canberra, Swinburne, Macquarie, and Buenos Aires. He is the author of *Robotomy* (1997) and *The Passion of Descartes* (2019), the latter about the French philosopher's quest to build an artificial replica of his dead daughter Francine.

**Sjoerd van Tuinen** is Associate Professor of Philosophy at Erasmus University Rotterdam, the Netherlands, and Co-founder of the Erasmus Institute for Public Knowledge. After his Ph.D. (Ghent 2009) on Leibniz and Deleuze, he has had visiting affiliations with universities in London, Berlin, Vienna, New York and Princeton. Driven by transdisciplinary problems across the arts and humanities, van Tuinen publishes on ecophilosophy, metaphysics, aesthetics, and the history of contemporary Continental philosophy. He is the editor of over a dozen books, including *Deleuze and The Fold. A Critical Reader* (Palgrave Macmillan, 2010), *Speculative Art Histories* (Edinburgh UP, 2017), *The Polemics of Ressentiment. Variations on Nietzsche* (Bloomsbury, 2018), and a series of theory books with V2_Institute for Unstable Media (Rotterdam). His current project is to finish a monograph entitled *Degrees of Perfection: A Mannerist Aesthetic.*

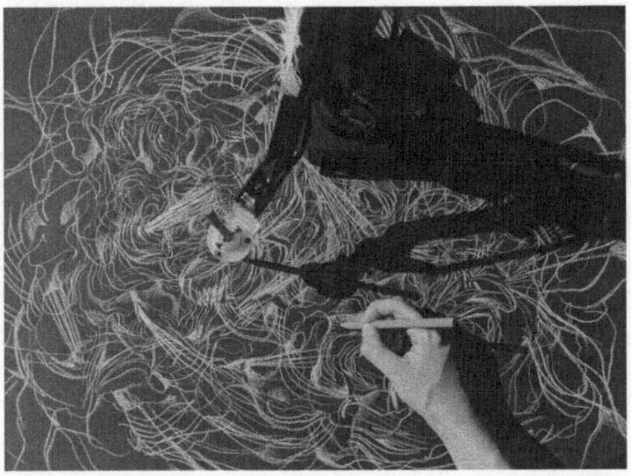

# PREFACE

yuk hui

# COSMOTECHNICS

When I was first invited by the editor of *Angelaki* to edit a special issue on cosmotechnics in 2017, I didn't pay much attention, since I didn't feel like editing a special issue dedicated to a concept that I had introduced only recently and was still to be fully developed. I only agreed to do so a year later, since it seemed to me that in the face of the global catastrophes of the present and the future, it is crucial to develop new conceptual tools to deal with the technological condition today. Pieter Lemmens was sympathetic with this project and was willing to participate in the editorial process – and in fact, he did a large part of the work. The concept of cosmotechnics is first of all a challenge to how technology has been understood in philosophy, anthropology and history in the twentieth century, which I will schematically present in terms of the following questioning.

In philosophy of technology, Martin Heidegger's famous 1949 Bremen lecture on the essence of technology, later published under the title *The Question Concerning Technology*, has been well received worldwide. In it Heidegger proposed that there is a rupture between what the ancient Greeks called *technē* and what he referred to as modern technology, for they differ in their essences. *Technē* has its essence in *poiesis*, i.e., bringing-forth, while modern technology or *enframing* [*Gestell*], sees everything as standing reserve or as resources to be exploited. We should ask ourselves, however, where the position of, say, ancient Indian technology, Chinese technology or Amazonian technology is in Heidegger's analysis? For sure, these technologies are not equivalent to modern technology, but can one assimilate or reduce them to Greek *technē*?

In anthropology of technology, the invention and use of tools (often covered by the terms labour or praxis) has been understood as the determining process behind hominization, convincingly demonstrated for instance by André Leroi-Gourhan. Technics has been interpreted by the latter as an extension of organs and an externalization of memory. In this interpretation, technology is anthropologically universal. This is not wrong in so far as such externalization and extension are considered as proceeding from what Leroi-Gourhan called a "technical tendency," but we still have to explain what he called "technical facts," which are different from region to region, and from culture to culture. What is embedded in these technical facts apart from a causal reduction to cultural difference, or even sometimes to contingency?

In history of technology, Joseph Needham raised a haunting question, namely, by asking why modern science and technology wasn't developed in China and India, while at the same time showing the large amount of scientific and technological development in China *before* the sixteenth century. Echoing Needham's inquiry, there have been significant inquiries on comparing technological development in different regions of the world in order to show that, for example, one particular region is more advanced in papermaking or metallurgy than another. However, this is a distortion of Needham's question, which in fact suggests that one cannot compare Chinese science and technology directly with that of the West, since they are based on different epistemologies and philosophies. In this sense, how can one re-articulate these differences?

These are some of the boundaries that the concept of cosmotechnics attempts to negotiate, since they all imply a universal concept of technology, which is in fact a residue of the desire of a particular kind of thinking. I gave a preliminary definition of cosmotechnics as unification between the cosmic order and the moral order through technical activities, in order to suggest that technology should be re-situated in a broader reality, which enables it and also constrains it. The detachment of technology from such a reality has resulted from the desire to be universalizing and to become the ground of everything. Such a desire is made possible by the history of colonization, modernization and globalization, which, being accompanied by its history of economic growth and military expansion, has given rise to a mono-technological culture in which modern technology becomes the principle productive force and largely determines the relation between human and non-human beings, human and cosmos, and nature and culture. The problems brought about by this mono-technological culture are leading to the exhaustion of resources and of life on earth and to the destruction of the environment, which are central to the discourse around the Anthropocene. It is also in this social and political context that it seems urgent to re-open the question of technology and the quest for a multiple cosmotechnics.

## acknowledgement

The editors would like to express their gratitude to all the contributing authors, to Dan Ross for the translations, to Sougwen Chung for the cover image and to Edwin Lo for editorial assistance. This special issue benefited from the support of the Cosmotechnics/Critical AI project funded by the City University of Hong Kong.

*Issue image: Sougwen Chung,* Artefact1 (Process)*, 2019, by permission.*

# INTRODUCTION

pieter lemmens

# COSMOTECHNICS AND THE ONTOLOGICAL TURN IN THE AGE OF THE ANTHROPOCENE

Heidegger's 1949 Bremen lecture on "The Question Concerning Technology" has been enormously influential in its conceptualization of the ontological essence of modern technology as being less that of the originary classical Greek notion of *technē*, understood as a bringing-forth of beings, than of what Heidegger called "enframing" [*Gestell*], understood as a mode of revealing of beings that provokes or challenges them forth exclusively as standing reserves [*Bestand*]. Under enframing, "nature" is being reduced to a gigantic reservoir of energy and material resources for economic and technological exploitation. While this image and critique of modern technology has been widely if critically accepted in the West, what has often been missed in those criticisms of its detail or trajectory is its applicability when extended to non-Western technological developments, both historically and projectively, for if these are not reducible to either *technē* or enframing in the Western sense, how can we articulate their unfolding and what could this question of technology prior to the Heideggerian formulation contribute to the current debate on the Anthropocene and the profound ecological and geological mutations it entails?

One current way forward in this context of the Anthropocene is a reappraisal of the idea of nature. For example, some representatives of the so-called "ontological turn" in contemporary anthropology such as Philippe Descola (*Beyond Nature*) and Eduardo Viveiros de Castro (*Cannibal Metaphysics* and *Relative Native*) have proposed an unravelling of the concept of nature to show that nature as it is experienced in non-European cultures cannot be equated to naturalism, which is a product of modernity, characterized by an opposition between nature and culture, in which nature is conceived as a universal ground that is common to all particular cultures arising from it while each having a different view or conception of that nature, which is adequately conceived though only by Western science. This naturalist framework of one universal nature – mono-naturalism – and many particular cultures – multiculturalism – is being confronted by these anthropologists with an alternative framework, principally derived from Amerindian animism, of a commonly shared culture – mono-culturalism – giving rise to many different natures – multi-naturalism – as well as with other ontologies, to wit totemism and analogism, that each understand the relation between nature and culture in yet other ways.

This mobilization of the non-modern can be seen as an attempt to reconceptualize the relation between the human and the non-human, and hence to go beyond the nature–culture dichotomy that restricts all visions to a parochial Western worldview. It is in the same spirit, but as a more pragmatically realist and political gesture, that this special issue of *Angelaki* addresses a parallel question for technology, which is whether it is also possible to conceive of multiple technologies, understood specifically in the sense of multiple cosmotechnics and if so what form this may take? It is our belief that it is not only the notion of "nature" that has to be challenged as the anthropologists did, but also the efficacy of any "return to nature" should be questioned, and that it is rather more urgent today to rethink the question of technology and of its possibilities for humanity's future existence and coexistence with non-human inhabitants on the planet.

Although the critique of modernism and Western naturalism from the viewpoint of non-modern and non-Western ontologies – such as practiced today by Descola and Viveiros de Castro but also others such as Tim Ingold, Eduardo Kohn and Robin Wall Kimmerer – is certainly worthwhile and fruitful for fostering the ontological imagination in our search for other, less destructive and less alienating modes of (co-)existence (cf. Weber), we share Clive Hamilton's reservations about the ultimate value of what he aptly calls "'going native' ontologically" for confronting the problems of the Anthropocene (106). The fact of the matter is that non-Western cultures are just as overwhelmed and disoriented by the Anthropocene as Western culture itself, although it may very well be the case that the Earth's indigenous populations are in some sense much better prepared for the impending breakdown of the capitalist world system than the well-pampered yet for the most part overly distressed, precarized and proletarianized inhabitants of the so-called "affluent societies" (Danowski and Viveiros de Castro 95–96), if only because they have already experienced the collapse of their cosmos (104). And the recent COVID-19 pandemic has demonstrated how fragile and arguably broken the global capitalist arrangement already is, with very dire prospects on the horizon.

More importantly still, the great majority of non-Western cultures have since long been enframed themselves by Western technology through colonization and imposed modernization such that the global technological condition has also become *their* destiny. Any "return" to native ontologies or conceptions of nature seems rather futile in this regard. All over the planet, a "second nature" that has been thoroughly affected by technology has come to replace the allegedly pristine "first nature" of yore. This is precisely the condition of the Anthropocene, where the technosphere – as the material condition of what Pierre Teilhard de Chardin and Vladimir Vernadsky have dubbed the noosphere in the 1920s – has come to overrule the biosphere. It is this development that is hardly if at all taken into account by the representatives of current anthropology's ontological turn, who generally fail to appreciate the decisive role of technology in anthropogenesis as well as in cosmogenesis, such as it has been emphasized in the work of the French paleoanthropologist André Leroi-Gourhan and more recently by the French philosopher and social critic Bernard Stiegler (*Technics and Time 1*). The ideas of these two authors have been particularly important for Hui in the development of his project of cosmotechnics.

Just as the ontological turn wants to do away with the opposition between nature and culture, the concept of cosmotechnics is designed to overcome modernity's opposition between nature and technology. Far from recommending a rehabilitation of pre-modern or non-modern ontologies or cosmologies though, the cosmotechnics project explicitly looks at the future and aims to be an imaginative and inventive discipline in search for *new* cosmotechnics, and that is to say a plurality of cosmotechnics for the age of the

Anthropocene, in which the different cultural spheres that exist across the globe are not called to retreat into their ancient cosmological structures – or what Peter Sloterdijk calls their traditional symbolic "immuno-spheres" (Sloterdijk and Heinrichs) – but to imagine and invent new practices of technological world-formation, taking *inspiration* from the ideas behind these structures, not in the least the moral or ethical ideas. It may indeed be "too late to exhume the corpse of Confucius" as Hamilton argues (107), but that does not mean it makes no sense at all to consult the Confucianist tradition – and by extension other traditions of thought beyond those of the West – for cues of how to think differently about "the unification between the moral order and the cosmic order through technical activities," as Hui's preliminary definition of cosmotechnics has it. A further objective of the cosmotechnics project, which it shares with Gilbert Simondon's genetic philosophy of technology, is to work towards a reconciliation of nature and technology.

When an anthropologist such as Descola develops an "anthropology of nature" (*Die Ökologie der Anderen*), it should also be possible to conceive of something like an "anthropology of technics," and if there are indeed multiple natures then it should also be possible to think of multiple technics that are different from each other, and not just functionally and aesthetically different – say at the level of technical artefacts – but different at the ontological and onto-historical level in the Heideggerian sense, or as Hui prefers: at the level of cosmology or histories thereof.

The concept of cosmotechnics was initially introduced in Hui's treatise *The Question Concerning Technology in China. An Essay in Cosmotechnics*, which understood itself first of all as a response to Heidegger's original question concerning technology but aiming to expand this question beyond Heidegger's still Eurocentric horizon and opening it up towards other, non-European cultures, such as Chinese culture which Hui, being Chinese himself, has taken as an example in his book. The concern that Hui shares with Heidegger is the fact that the Western technological trajectory has come to dominate nearly all other cultural spheres around the globe and has indeed become a planetary force – the gigantic, monolithic force foremost responsible for the arrival of the Anthropocene as the age of generalized psychic, social, cultural, economic and ecological deterioration or "entropization," as Bernard Stiegler has put it, characterizing the Anthropocene as the Entropocene (*Automatic Society*).

According to cosmotechnics all technics are fundamentally cosmological and all cosmologies are fundamentally technical, and what it underlines is that technology is not a universal, transcultural category but essentially dependent in its unfolding on a non-technical cosmological factor that is culture-specific. As such it aims to rethink and re-orient the question concerning technology in terms of the question concerning technodiversity. Its emphasis on multiple cosmotechnics has to be distinguished from so-called multiculturalism, which is fundamentally a reactionary politics of cultural identity. The multiplicity of cosmotechnics implies multiple epistemologies and epistemes which can contribute to reflection on the development of technologies, opening up a discourse on technodiversity for overcoming the current homogenization and planetarization of enframing actualized in the Anthropocene. Hui, again, uses China as an example to demonstrate that it is not only possible but also necessary to elaborate on culture-specific histories of cosmotechnics, and to see how these different understandings can be reconsidered and reevaluated so as to re-orient and re-imagine current technological globalization with its nearly complete obliteration of technodiversity from a cosmotechnical perspective.

The ultimate concern of the cosmotechnics project, which is further developed by Hui in *Recursivity and Contingency* and in *Art and Cosmotechnics* (forthcoming in 2020), is to show the necessity of overcoming the current global dominance of Western or promethean technology, re-interpreted as one particular form of cosmotechnics (i.e., that of capitalism). Its ultimate objective is to encourage another, non-Eurocentric way of thinking about technology in order to bring about what Hui calls a bifurcation or rather fragmentation of World History such that it may open the future toward a plurality of heterogeneous technological trajectories, each driven by a different cosmotechnical imaginary. As such it projects a profound "reframing" of the current planetary enframing so as to "recosmicize the Earth" (Augustin Berque) and doing so in a variety of locally specific ways.

In this special issue, ten authors explore various dimensions of the cosmotechnics proposal, some in closer dialogue with Hui's work, others by relating to it more tangentially and taking it into other, sometimes very different, directions. I have to confine myself here to an extremely concise exposition of the authors' contributions.

*Pieter Lemmens* interprets Hui's cosmotechnics in terms of a "critical synthesis" of Heidegger's and Stiegler's views of technology which updates them for the emerging post-Eurocentric epoch. *Marco Pavanini* situates cosmotechnics within an anthropotechnological and culture-historical perspective and shows that it, as an inevitably comparative discipline, requires reflexive awareness of the various "cultural techniques" that condition any culture's self-understanding so as to guard itself against ethnocentrism, suggesting the cultural techniques approach developed among others by Thomas Macho to be the proper "organon" of cosmotechnics. *Andrés Vaccari* emphasizes the affinity of Hui's cosmotechnics with that of Gilbert Simondon and then mobilizes both to offer a scathing critique of the hegemonic cosmotechnics of today's hypercapitalist conjuncture, that of transhumanism/singularitarianism, interpreted as a corporate, neosubstantivist ideology serving the agenda of a rabid neoliberalism intent on totally controlling and colonizing the planet. *Yuk Hui* for his part argues that cosmotechnics can help to overcome the totalizing tendency of the currently dominant cybernetic paradigm structuring the relation between nature and technology, or between ecology and machines, by offering a more pluralistic, historical and regional understanding of this relation, enabling what he calls a genuine "political ecology of machines." *Bernard Stiegler*, in an article that relates Hui's concept of technodiversity to his own notion of noodiversity, argues for the necessity to develop an entirely new theoretical computer science with a view to countering the entropic tendencies inherent in the global digital milieus of contemporary "platform capitalism" so as to rethink the relation between the nervous and the logical and resurrect a new space for the exercise of reason, which is now in the process of being dissolved in a digitally automated understanding, itself completely reduced to calculation. *Jason Tuckwell* takes Hui's idea of cosmotechnics as starting point for a critique of technicity conceived as a uniquely human characteristic, a viewpoint derived from Aristotle, arguing instead that technically mediated agency is constitutive for all species and showing that this articulates nicely with the non-universalist conception of technics embraced by Hui and allows for the possibility of non-humans also having a cosmotechnics. *Sjoerd van Tuinen*, in a historically informed reflection on the possible future(s) of artificial intelligence, more or less presents Leibniz as a cosmotechnic thinker of technological and noetic diversity avant la lettre, contrasting his proto-cybernetic idea of reason as distributive composition and unnatural common sense with Hegel's collectivization of an artificial yet self-naturalizing good sense. *Clive Hamilton* argues that exploring ancient cosmological frameworks for conceiving new future cosmotechnics in the plural may be wrongheaded since they all originate from the definitively past epoch of the Holocene, whilst the development of what he calls a "fifth ontology" beyond those four currently held by humans on the planet (as distinguished by Philippe Descola) will necessarily originate from our progressive interactions with the entirely unprecedented Anthropocene Earth. *Frédéric Neyrat* responds to the cosmotechnics proposal affirmatively yet critically, from his own quite unique "Alienocene" perspective, by presenting "Afrofuturism" as an alternative, non-imperialist, non-colonialist and non-exploitative cosmotechnics countering the violent, acosmic cosmotechnics of capitalism. In this Afrofuturist cosmotechnics, the cosmos is not limited to our terrestrial abode though but designates the entire universe as a trajectory extending all the way to its anarchic origin in the so-called Big Bang, reminding us of the ultimate futility of any technological will-to-power. *Peter Skafish* finally relates to the cosmotechnics proposal by reflecting, from a more anthropologically or rather anthropotechnologically informed approach and in dialogue with anthropology's contemporary ontological turn, on the body and on "techniques of the body" (a term coined by Marcel Mauss) from a pluri-ontological and "polyrealist" perspective, showing how the body and its technicity manifest in very different ways according to its ontological as well as cosmological embedding, illustrating this in particular with reference to Amerindian shamanism and – most extensively – traditional Chinese medicine.

It will be obvious from this all too brief preview that cosmotechnics is a concept that opens up many possible avenues for philosophical reflection, offering a fresh perspective for rethinking the human–nature–technology nexus in multiple ways. We hope that this special issue marks the beginning of the development of cosmotechnical thinking as a new direction for the philosophy of technology, that also allows for an enriched dialogue with other disciplines such as ethnology, anthropology, cultural history, social and political science, and not in the least also the arts. We thus hope that it may inspire future critical thinking about the human condition in the context of the Anthropocene and may support imagining a new technological modus vivendi on this planet or rather a multiplicity thereof, and we also hope, finally, that the cosmotechnics project will be taken up by different cultures and different localities on the planet.

## bibliography

Danowski, Déborah, and Eduardo Viveiros de Castro. *The Ends of the World*. Trans. Rodrigo Guimaraes Nunes. Cambridge: Polity, 2017. Print.

Descola, Philippe. *Beyond Nature and Culture*. Trans. Janet Lloyd. Chicago: U of Chicago P, 2013. Print.

Descola, Philippe. *Die Ökologie der Anderen*. Trans. Eva Moldenhauer. Berlin: Matthes, 2014. Print.

Hamilton, Clive. *Defiant Earth. The Fate of Humans in the Anthropocene*. Cambridge: Polity, 2017. Print.

Heidegger, Martin. *The Question Concerning Technology and Other Essays*. Trans. William Lovitt. New York: Harper, 1977. Print.

Hui, Yuk. *Art and Cosmotechnics*. Minneapolis: U of Minnesota P. Forthcoming.

Hui, Yuk. *The Question Concerning Technology in China. An Essay in Cosmotechnics*. Falmouth: Urbanomic, 2016. Print.

Hui, Yuk. *Recursivity and Contingency*. London: Rowman, 2019. Print.

Ingold, Tim. *The Perception of the Environment. Essays on Livelihood, Dwelling and Skill*. London: Routledge, 2011. Print.

Kohn, Eduardo. *How Forests Think. Toward an Anthropology Beyond the Human*. Berkeley: U of California P, 2013. Print.

Leroi-Gourhan, André. *Gesture and Speech*. Trans. Anna Bostock-Berger. Cambridge, MA: MIT P, 1993. Print.

Sloterdijk, Peter, and Hans-Jürgen Heinrichs. *Neither Sun Nor Death*. Trans. Steve Corcoran. Cambridge, MA: Semiotext(e), 2011. Print.

Stiegler, Bernard. *Automatic Society Vol. 1. The Future of Work*. Trans. Daniel Ross. Cambridge: Polity, 2016. Print.

Stiegler, Bernard. *Technics and Time 1. The Fault of Epimetheus*. Trans. Richard Beardsworth and George Collins. Stanford: Stanford UP, 1998. Print.

Viveiros de Castro, Eduardo. *Cannibal Metaphysics*. Trans. Peter Skafish. Minneapolis: Univocal, 2014. Print.

Viveiros de Castro, Eduardo. *The Relative Native. Essays on Indigenous Conceptual Worlds*. Chicago: Hau, 2015. Print.

Wall Kimmerer, Robin. *Braiding Sweetgrass. Indigenous Wisdom, Scientific Knowledge and the Teachings of Plants*. Minneapolis: Milkweed, 2015. Print.

Weber, Andreas. *Indigenialität*. Berlin: Nicolai, 2018. Print.

pieter lemmens

# OTHER TURNINGS

## yuk hui's pluralist cosmotechnics in between heidegger's ontological and stiegler's organological understanding of technology

In this article I will present Hui's cosmotechnical understanding of technology as a kind of "critical synthesis" of Heidegger's and Stiegler's views on technology,[2] showing that, on the one hand, it acknowledges Stiegler's insight into the constitutive technicity – universally valid – of human existence and what Heidegger called "world-formation" [*Weltbildung*], while on the other it pays tribute to Heidegger's thesis that all technical evolution is always already conditioned by a non-technical factor, albeit for Hui not a singular *onto*logical factor but a plurality of always particular *cosmo*logical factors.

### introduction

Yuk Hui's *The Question Concerning Technology in China. An Essay on Cosmotechnics* is a rare yet timely attempt in these days of narrow-minded empirically oriented and case-study saturated philosophy of technology to re-consider the question of technology as a truly philosophical question in the most profound and urgent manner, such as it was first posed by Martin Heidegger in 1949 in his notorious Bremen lecture on *The Question Concerning Technology*, where he famously asked about the *essence* of technology, which he deemed to be nothing techno-logical, as is well known, but ontological or onto-historical – as having to do with what his own philosophical questioning had been after from the very beginning and that was: Being – and, later, the *history* of Being.[1]

Hui's project of cosmotechnics seems to a large extent inspired by Heidegger's ontological or onto-historical understanding of technology as an unfolding essence or *Wesung* overdetermining all concrete technological invention and innovation, replacing it though with a plurality of culture-specific cosmological or cosmotechnical trajectories that allow for the thinking of a profound technodiversity which has no place within Heidegger's thought of a singular essence of technology. Yet it is also decisively informed by Bernard Stiegler's critique of Heidegger's neglect of the constitutive nature of technology in its concrete factuality for any ontological framework.

That is to say: Hui's cosmotechnics professes *both* the technicity of the cosmic (understood as the world) *and* the "cosmicity" of technics. Employing a distinction made by Peter Sloterdijk, I will try to demonstrate how it thus combines Stiegler's emphasis on the "horizontal" dimension of technology as technical exteriorization with Heidegger's claim about the "vertical," in the sense of "spiritual" or "cultural," determination of technology, pluralizing both these dimensions. In doing so, it opens up these fundamental philosophies of technology to the debate, increasingly urgent in our dire time of the Anthropocene, with contemporary anthropology's so-called "ontological turn."[3]

## the planetary reign of western technology

Heidegger questioned technology, i.e., its essence, not for the sake of finding a right definition or a correct representation of it but to prepare a *free relationship* to it (*Technology* 3). And this free relationship would be attained only when humans stopped focusing – "ontically" – on technical objects and systems or even taking a (critical, evaluative, moralizing, etc.) stance toward them and would instead open themselves to technology's *essence*; and this first of all meant to recognize that this essence resided in nothing less than humanity's overall relation to beings *as rooted in* a specific *understanding of Being* – an understanding of Being that they principally inherited, he argued, from the ancient Greeks and that ruled with ever more insistence and exclusivity as a "destiny" [*Geschick*] over all their encounters with beings or "nature" generally.

Yet opening up to the essence of technology and becoming aware of being destined by it in their *own* essence – in their very way of being – could "awaken" in humans a "remembrance" of their true, ontological essence of *belonging* to Being or *being open* to Being and thereby grant them the possibility of entering into – what Heidegger considered to be – a more original relation to Being and thus to experience a more inceptual truth of Being (*Technology* 28), one that was first "revealed" to them at the Greek inception but almost immediately "forgotten." This renewed experience of the truth of Being would reconnect humans with their original openness to Being or what I would like to call their *ontological freedom* here. It is indeed this *free relationship* to the essence of technology *as Being* which Heidegger's thinking attempted to liberate, and only this freeing oneself to the freedom of Being – as a "re-engagement" of humans with their *originary* ontological freedom granted by Being – would allow in his view a genuine "overcoming" or "turning" of the dominance of modern technology over our lives.

Heidegger's questioning of technology in the Bremen lectures was first of all meant as an attempt to render our ingrained and self-evident technical (as well as scientific) relation to beings "worthy of questioning" [*fragwürdig*] in a radical way – by "dignifying" [*würdigen*] its essence as being of the highest importance for philosophical thinking and human existence more generally (*Insight* 57). It is worth repeating here that this essence, for Heidegger, was itself not techno-logical but ontological or even more precisely: aletheio-logical, i.e., a mode of revealing of beings – *aletheuein* in Greek – in accordance with a certain unconcealment or truth of Being. The question concerning technology for Heidegger was ultimately the question concerning Being or rather the truth of Being and this made it *eo ipso* the philosophical question par excellence.

It is clear that when Heidegger spoke about the human in his technology essay he first of all had "Western man" in mind and that his analysis of technology concerned Western technology, in particular *modern* Western technology, but already in the 1930s he emphasized that this modern Western technology had become a "planetary" phenomenon (*Mindfulness* 13; *Geschichte* 74) and was adopted by practically all non-Western, i.e., all non-European cultures as well. And in his "Time and Being" lecture from 1962 the late Heidegger enigmatically stated, putatively anticipating the contemporary notion of the "technosphere" (Haff), that:

> [n]ow that modern technology has arranged its expansion and rule over the whole earth, it is not just the sputniks and their by-

> products that are circling around our planet; it is rather Being as presencing in the sense of calculable material that claims all the inhabitants of the earth in a uniform manner without the inhabitants of the non-European continents explicitly knowing this or even being able or wanting to know of the origin of this determination of Being. (Heidegger, *On Time* 7)

The whole planet had now been seized *uniformly* by the reign of enframing [*Gestell*], an imperious, extractive and calculative relation to beings of European provenance increasingly overruling all other "cultural" origins, without these other cultures explicitly being aware of this provenance. And also, we may assume, through the suppression of these other origins, resulting in the techno-cultural uniformization or homogenization of the planet. What the late Heidegger feared yet considered to be inescapable is what he referred to – in his conversation from 1954 with the Japanese scholar Tomio Tezuka – as "the complete Europeanization of the earth and of man" (*Language* 15), unfolding as the planetarization of modern technology.

It is *this* concern that also motivates Hui's cosmotechnics project, which aims to re-open the question concerning technology, first of all in regard to the Chinese situation, in light of the danger of modern technology attaining complete planetary dominance and absorbing all cultural spheres – i.e., the whole *ethnosphere* to speak with the American anthropologist Wade Davis (2) – within its "homogeneous becoming" (Hui, *Cosmotechnics* 12), thereby effectively annihilating all cultural difference through a nihilist process of technological indifferentiation. Yet, while Hui fully subscribes to Heidegger's diagnosis of the planetarization of modern technology and profoundly shares his concern about the danger of this development – this danger having been intensified to an extreme currently in the context of the Anthropocene – he challenges Heidegger's assumption that there is, ultimately, only one kind of technology and argues that there are other kinds as well.

And with his notion of technology he seems to have something similar in mind, in my opinion, as what Heidegger called the essence of technology, although it is also quite different considering the fact that he proposes to talk about cosmotechnics instead of some "essence" of technology, and indeed of a *plurality* of cosmotechnics instead of the unitary, monolithic essence of technology identified by Heidegger,[4] although the latter distinguished Greek *techne* from *die moderne Technik* of course, thereby acknowledging a radical cesura within the historical unfolding of technology yet considering this cesura within the continuity of the history of *Being* as the history of a decline from the *inceptive techne* toward its *conclusive* manifestation in *die moderne Technik*, understood as a progressive and cumulative "decline from the inception" or "perversion of the inception" (*Being* 9) – and this is precisely what his talk of the danger as the gathering [*Ge-*] "entrapping" [*fahr*; being derived from *fara*, translated as *Nachstellen*] by the forgetting of Being refers to (*Insight* 50).

## singular or plural technics?

As Heidegger puts it in the most straightforward manner at the beginning of his 1943 lecture course on Heraclitus: "There is only an Occidental technology. It is the consequence of 'Philosophy' and nothing else" (*Heraclitus* 3). Philosophy for Heidegger is the ground, the *only* ground, out of which technology has arisen and philosophy is an exclusively Western affair (3). Nonetheless, this Western technology, together with "the sciences," which for Heidegger share the same essence, has now come to dominate the whole planet and become "international," as he writes in an entry in one of his *Black Notebooks* that is also quoted by Hui (Heidegger, *Anmerkungen* 59–60; Hui, *Cosmotechnics* 4–5). This implies, and who could deny this given its global dominance, that Western technology "is something detachable from its cultural source" (Hui, *Cosmotechnics* 5).

However, because of its international presence and penetration into practically all cultures around the globe nowadays the erroneous idea has come to prevail that technology is a *universal* phenomenon and thus independent of any particular culture – indeed transcultural. Heidegger instead emphasized its unique European origin insofar as it issued from Western philosophical thinking, as can be gathered very clearly for instance from this remark at the end of his first Freiburg lecture on the "Basic Principles of Thinking" from 1957:

> [...] if we call upon the fact that in our age everywhere upon the earth a uniform manner of thinking achieves world-historical dominance, then we must just as decisively hold in view that this uniform thinking is only the form, leveled down and rendered useful, of that historical formation of thought that we name the Western-European, the dispensational singularity of which we scarcely even experience and seldom enough acknowledge. (*Insight* 89–90)

It is indeed true that Heidegger never recognized any other technology besides that of the West, which originated in Greek *techne*, then evolved into modern technology in the seventeenth century and finally culminated in our age in the single, homogeneous and uniform, planetarized "gigantic force" – as Hui often refers to it (*Cosmotechnics* 239, 243, 297) – of enframing, or with an earlier term: machination [*Machenschaft*], turning the whole of the planet into a standing reserve [*Bestand*] (Hui, "Cosmopolitics" 4). Again, we should emphasize that technology for Heidegger here refers to the *ontological–aletheiological* essence of technology, i.e., to a mode of revealing of beings. Whether this is true for Hui or not (and we will address this question later), he understands what Heidegger called enframing as only *one* kind of cosmotechnics, that of today's global capitalism (Hui, *Cosmotechnics* 299).

Hui in fact never explicitly talks about the essence of technology in relation to his notion of cosmotechnics, yet when he indicates that his attempt to re-open the question concerning technology in terms of a plural cosmotechnics is intended as a dialogue with "Heidegger's concept of technics" (*Cosmotechnics* 7), he writes that what he has in mind is "the philosophical concept of technics" and then characterizes this philosophical concept as an "ontological category," arguing that it comes to the fore with the birth of philosophy (10). So although it seems that the essence of all technics for Hui is *cosmo*logical (or cosmotechnical), it also signifies something *onto*logical (and we will have to examine what the difference exactly amounts to). And it is also *plural* given the plurality of cosmological origins of different technics, in contrast to the *singular*, Greek origin of technics proposed by Heidegger. Hui explicitly hypothesizes fundamentally different *kinds* of technics, and that is to say different not simply in a functional and aesthetic sense, but ontologically and cosmologically different (xiii). And instead of the singular "turning" envisioned by Heidegger, he entertains the possibility of a pluralistic "bifurcation" or rather "fragmentation" (*Recursivity* 34, 257, 261–64) toward different technological futures ensuing from different, yet to be imagined future cosmotechnics ("Cosmopolitics" 7), all the while acknowledging the current hegemony of Western – sometimes also referred to as "Promethean" (*Cosmotechnics* 12–14, 29, 33, 37, 196, 229) – cosmotechnics.

So like Heidegger, Hui argues that technology *as we know it* is not universal but indeed originates with the emergence of philosophy in ancient Greece, the birthplace of the West. It is thus specific to Western culture and does not exist in China since the latter never developed a philosophical thinking like that of the Greeks (Hui, *Cosmotechnics* 16). Nonetheless, again, this Western concept of technics is in the process of totally overpowering Chinese culture at the moment (as well as practically all other cultures on the planet of course), to the point of completely obliterating its own cultural origins. To counter this process, Hui argues, the Chinese should not and cannot simply follow the Heideggerian trajectory of "returning" to the Greek origin of technics, since there never "was" such an origin in China. They should instead "re-connect" – in a sense not dissimilar to Heidegger's "anamnesis" of the Greek inception – with their own techno-cultural origin, which has been very different from that of the West and which he aims to uncover precisely in terms of a (Chinese) cosmotechnics.

In the first part of his cosmotechnics book, then, Hui attempts to distill something like the philosophical–ontological concept of technology sensu Heidegger, albeit an implicit one, for the Chinese cultural sphere, from an in-depth reading of the Chinese tradition of thinking starting from its Daoist and Confucianist beginnings, focusing on the basic concepts of *Qi* (器, utensil, to be distinguished from 氣, literally gas, energy) and *Dao*. And he does so in the form of a cosmotechnics, suggesting on a few occasions in the book that this may also be what Heidegger was after in his search for the original essence of technology as *techne* in ancient Greece (Hui, *Cosmotechnics* 77, 79, 116, 191). This is because technology as an ontological category, he contends, "must be interrogated in relation to a larger configuration, a 'cosmology' proper to the culture from which it emerged" (10).

The question that immediately comes up here is what the difference might be exactly between an ontological–aletheiological concept of technology sensu Heidegger and the cosmological concept Hui seems to propose, understood as the larger cosmological configuration from which he claims that ontological concepts of technics spring? Why is the cosmological more encompassing than the ontological, and what is the difference actually between an understanding of *Being* and an understanding of the *cosmos* insofar as these are taken to overdetermine and/or condition conceptions of technics and concrete technical development? And what exactly is the difference between a cosmo*logy* and a cosmo*technics*? And even more basically: what does Hui understand by a cosmology in the first place? What is a cosmic order? And how does he think the *relation* between cosmology and ontology? Or between cosmotechnics and ontology? And what does he mean when he talks, with a word also used frequently by Heidegger in relation to the essence of technology, about the *spirit* of the ancient Greek – or Chinese – cosmotechnics (Hui, *Cosmotechnics* 79, 164, 168). And most importantly for what will follow: how does he conceive of the relation between ontology and technology from his cosmotechnical perspective?

In this article I simply want to explore these questions, especially the last one, against the backdrop of, first, Heidegger's ontological or rather onto-historical understanding of technology as I've briefly laid it out here above and, second, Stiegler's "onticized" and "anthropologized" – or we might say "onto-anthropological" (Sloterdijk, *Not Saved* 98) – *organological* understanding of technology, which has quite some affinities, as I will show, with Peter Sloterdijk's onto-anthropological understanding, which is also developed as a critical response to Heidegger's views. It seems to me that Hui's notion of cosmotechnics also represents an effort, at least implicitly, to "synthesize" these two opposing or in any case conflicting yet also deeply related conceptions of technology, and also an attempt to pluralize them both. The project of cosmotechnics proposed by Hui also seems to address the shortcomings of both the Heideggerian and the Stieglerian understanding of technology, the former for its lack of thinking technics concretely, the latter for lacking a (multi)cultural and/or (pluri-)ontological perspective on technical development, and both for being too "Eurocentric" or "Western-centric" (in this critique he also invokes Gilbert Simondon's views on technology but I cannot go into that in this article).

## being, cosmos and technology in heidegger, stiegler and hui

Describing the difference between Heidegger's and Stiegler's understanding of technology in the most simplifying way we could say that whilst Heidegger asserts the ontological essence and origin of technology, Stiegler on the contrary claims the technological essence and origin of all ontology. Hui for his part seems to assume something of a middle position as I hope to show in this article. Since Heidegger and Stiegler have a different understanding of technology, they also have a different understanding of the *danger* of technology. For Heidegger, as we've seen, this danger consists ultimately in the forgetting of Being and an exclusive focus on beings, which is what metaphysics amounts to in his view. Enframing as the essence of modern technology is then the completion of metaphysics, instituting the *supreme* danger of a *total* forgetfulness of Being and a complete loss of the ontological freedom of Dasein, which then comes to understand even itself solely in terms of a standing reserve (Heidegger, *Technology* 27), to be optimized technologically for functioning within its technical environments, as today's so-called transhumanists aspire to. For Stiegler, on the contrary, we could say that the danger of technology consists in the fact that it *itself* is constantly forgotten, not in its ontological essence but in the ontic sense of concrete technical artefacts as they constitute, as he contends, every understanding of Being that Dasein might be capable of, and condition all of its possibilities. For Stiegler, metaphysical thinking consists precisely in the forgetting of technology in the sense of a forgetting or neglecting by the human of being conditioned – originally – by the technical artefact, and that is to say by a process of technical exteriorization or what he has come to call *exosomatization* recently, adopting a term from the Polish-American biologist Alfred Lotka. I will come back to this in more detail below.

It appears to me that one can recognize this contrast between the Heideggerian and the Stieglerian view of technology in the "antinomy" presented by Hui regarding the universality of technology. Whilst his cosmotechnics project definitely questions this universality, it nonetheless *does* acknowledge the existence of an anthropologically universal aspect of technology. With the French paleoanthropologist André Leroi-Gourhan, who has theorized human evolution as proceeding from a process of technical exteriorization and the extension of organs into technical artefacts, he distinguishes between a transcultural and thus anthropologically universal *technical tendency* and the particular ways in which this tendency sediments in different cultures into a variety of *technical facts*. Whilst Leroi-Gourhan – and implicitly also Stiegler, we may add – explains this particularization and diversification of the universal technical tendency into various technical facticities or specificities through the influence of the factual circumstances of the local *milieu* in which the technical tendency operates, Hui contends that this cannot be the whole story and suggests that another, non-universal but particular explanatory factor needs to be taken into account and this is precisely the cosmological understanding or cosmological setting characteristic of the culture in question, which is why he claims that technics is always *cosmo*technics (*Cosmotechnics* 19).

Although Hui frames these alternative views in the form of a Kantian antinomy as contradictory theses concerning the true nature of technology, one could also simply distinguish two *dimensions* of technology: a universal one rooted in the "nature" of the human species and a non-universal one rooted in an always local cosmology, which is in fact also how he rephrases it in his most recent book *Recursivity and Contingency* (265–66). Now it is obvious that the universal dimension of technology as technical exteriorization also characterizes Stiegler's anthropological understanding of technology, who follows Leroi-Gourhan here, but adds to it that this process also explains the coming into being, at a very particular stage of its unfolding, of Dasein as an ek-sistent being open to Being, capable of understanding the Being of beings and of forming worlds, i.e., of shaping a cosmos. As Stiegler writes: "we access the cosmos as cosmos on the basis of hypomnesic tertiary retentions [ = technical artefacts; P.L.] in all their forms, from the shaman's instruments to Herschel's telescope" (*Nanjing* 42). As for the ontological capacity of humans, for Stiegler it is "the technico-historic fact of writing," i.e., the emergence of alphabetic writing technology – as "orthographics" allowing for the exact recording of temporal experience – that grants humanity the ontological freedom which Heidegger considers as a pure gift of Being, and that inaugurates the history of Being at the Greek inception (*Disorientation* 34, cf. 12).

The non-universal, particular dimension that Hui determines as cosmological or rather cosmotechnical and also identifies as culture-specific and characterized by a certain "spirit," seems to me close to what Heidegger aimed at, albeit only focused on the West, when he tried to think *techne* as the essence and also the "essential genealogy" destining the history of Being as an insistent "claim" [*Anspruch*] originating from the Greek inception and culminating in the claim of enframing that today reigns at the planetary level (*Insight* 63). It is this, for Heidegger irreducibly ontological or rather onto-historical "essential provenance" [*Wesensherkunft*] of our technological civilization – as the "veiled inceptual essence of being" (63) – that Hui tries to rethink, or so it seems, and also to pluralize, for cultural origins different from the West (i.e., Greece), in terms of primal cosmologies or cosmotechnics. The latter goes "beyond cosmologies" (Hui, *Cosmotechnics* 309), he writes, in that it acknowledges (and this is again Stieglerian we can say) the role of technics and not just of language and cultural practices we may assume, in the constitution of a cosmos (24), but most importantly for Hui is that all "technics is both driven by and constrained by cosmological *thinking*" (217; my emphasis).

Similarly for Heidegger all concrete technological innovation is always already driven – destined that is – by a technological understanding or revealing of beings, i.e., a kind of *thinking*, that has its root in the inceptual revelation, *increasingly forgotten*, of the truth of Being – *aletheia* – in ancient Greece and that will continue to exercise its grip on Dasein until, and this is crucial for understanding the difference between Heidegger and Hui, it overcomes this forgetting and remembers its *originary* "innermost indestructible belongingness to being" (Heidegger, *Technology* 32), which re-turns it to its originary ontological freedom and grants it access to the openness of Being. This "turning" would inaugurate what Heidegger called the other inception [*andere Anfang*], i.e., a new beginning, beyond metaphysics as the history of the forgetting of Being as it has culminated in the planetary reign of enframing, and toward the "event of appropriation" [*Ereignis*] (*Insight* 117), the event that "appropriates man and Being to their essential togetherness" (*Discourse* 38) and through which Dasein would truly assume – for the very first time – its more originary essence as "the founder and the preserver of the truth of being" (*Questions* 181) and become what Heidegger called the "shepherd of Being" [*Hirt des Seins*] in the letter "On Humanism" (*Writings* 234), instead of continuing to pursue – in total blindness to Being – the path of technological domination and exploitation of the Earth as the alleged "lord of beings" [*Herr des Seienden*] (245).

We have to emphasize here that this turning into the event of appropriation, "in which man and Being are delivered over to each other" (Heidegger, *Discourse* 36), entails nothing less, for Heidegger, than a complete "transformation of humanity itself" (*Questions* 181), a radical change in the very essence of the human resulting from a "change in being" (*Insight* 65). This change is characterized in the letter "On Humanism" as that of the transition from the *homo animalis* to the *homo humanus* (Heidegger, *Writings* 254), where the former refers to the *animal rationale* of metaphysics, which thinks the humanity of the human from its animalistic origin, i.e., as living organism endowed with an additional capacity such as reason, spirit or soul able to determine and categorize beings (227), while the latter thinks the humanity of the human "from nearness to Being," i.e., from "its provenance from the truth of Being," *which Heidegger deems more originary* (245).

We will see below that Stiegler contests this originary status of Being, assigning the origin not to rational animality though, as metaphysics does, but to technicized animality, attempting to rethink existential ontology and the history of being organologically and that is to say in terms of techno-organicity. From a Heideggerian viewpoint though Stiegler's thinking nonetheless appears to remain locked within the horizon of *animalitas* and indeed he understands the human with Georges Canguilhem as a "technical form of life" (*Automatic* 12, 223) or indeed an "exosomatic form of life" (*Neganthropocene* 249), "naturalizing" or "de-transcendentalizing" the rationality of the *animal rationale* as it were through technics.[5] Hui for his part thinks that the sheer process of technical exteriorization and the becoming organological of the human animal sensu Stiegler and

Leroi-Gourhan is *not* enough to account for the way in which technology evolves and suggests as an additional determinant the influence of a cosmological a priori.

The change in being from enframing to the event of appropriation and the consequent change of the essence of humanity intended by Heidegger cannot be enforced by humans since humanity is not the "master of being" (*Writings* 245). For Heidegger human action by itself is impotent regarding the essence of technology holding sway over its being. The only possibility of "overcoming" its reign or rather of "converging" [*verwinden*] it into its hidden essence as the event of appropriation is to *open* oneself to it, to "become attentive" to it (Heidegger, *Insight* 66), to *hear* its call, to *experience* its imperative, to make it worthy of *questioning* and then try to *think* it. Focusing, however critically, on concrete technological developments and their effects on human life and society is of no use whatsoever according to Heidegger; indeed, this only further consolidates our implication in enframing. It is from this resolutely non-ontic understanding of technology and consequently "anti-activist" (but equally "anti-passivist") stance toward technology, that Heidegger's well-known dictum that "only a god can save us" as well as his later discourse on releasement and non-willing should in my view be understood.

## the "verticality" of technology as being in heidegger

What is most important though, and this is something that Stiegler emphatically resists but Hui seems willing to account for in some sense, is that for Heidegger Being as the essence of technology and as that "[w]hat essences of danger" (*Insight* 59) is more originary and more "real" in fact than any concrete technology and principally beyond the realm of the ontic as such.[6] And for Heidegger this also means that it is only in Being, as *wholly different* from beings, that the "saving power" resides, *never* in beings (simply because the claim of technology issues from Being and not from beings). This is something that cannot be emphasized strong enough in my view. When Heidegger talks about the turning, the new beginning or the other inception, he always insists that this event, which does not "happen" *in* the world but *with* the world, is something that can only happen "suddenly," i.e., in a flash and "without mediation" (*Technology* 44). Being as the omnipresent essence of (the danger of) technology has nothing equal to it, he writes, thus it cannot be brought about by any being whatsoever and neither does it bring about any being itself (44). Being therefore does not belong at all to the realm of causality, i.e., to "effectiveness" [*Wirksamkeit*], despite the fact that it "determines" the presencing of beings as such – i.e., as effectiveness – in its modality of enframing (Heidegger, *Insight* 39). When the turning "happens," it does so "abrupt" and "steep," "out of its own essence of concealedness" (Heidegger, *Technology* 44). It can only come to pass as the "sudden self-lightning of being," in which "the truth of Being flashes" and thereby "the essence of Being clears and lights itself up" (44). And it "presences" [*west*] only as a "favor" [*Gunst*] within the danger, a favor thoroughly immune to any enforcing, for which humans can therefore only prepare and try to become receptive in their very Being.

This "steepness" or what I would like to call the "verticality" – and in a proper sense also the "divine" nature[7] – of the turning and the event of appropriation, and the absolute receptivity of Dasein it presupposes, is absolutely key to understanding Heidegger's view and evaluation of technology's onto-dynamic essence and how it fundamentally differs from any anthropological understanding of technology such as that from Stiegler, which assigns primacy to the "horizontal" process of technical exteriorization and considers any relation to Being as conditioned by it. This verticality is also emphasized by Peter Sloterdijk in a very interesting essay on what he calls the "kinetic features" of Heidegger's existential and onto-historical thinking. According to Sloterdijk one can distinguish three universal and fundamental kinetic features or motions in the movement of being within which human existence is caught: first the *fallenness*, thrownness or plungedness of existence (from an originary inception) into the drift of everyday preoccupations, then *experience* as the exploratory, experimental, discovering and inventive, horizontal and cumulative movement of "coming-to-the-world" [*Zur-Welt-kommen*], and finally *reversal* or the revolutionary *turning* of Being into a renewed (and more originary "appropriated") originarity, or a new, yet more inceptual inception (*Not Saved* 13–15).

In Heidegger's ontokinetics, though, the second movement of experience hardly plays a role and is in fact totally downplayed in favor of an almost exclusive focus on the first and third movements. In fact, the horizontal plane of experience – to be identified here with technical invention and scientific research or in Stiegler's terminology: technical exteriorization – has no constitutive role at all and is fully determined and characterized *in advance* by the movement of falling from the inceptive revelation of being. One sometimes has the impression that concrete technologies for Heidegger simply emerge as a result from Dasein's insistence in the forgetting of being and obeyance to the imperative of enframing, as nothing but "continuations of the plunge in the horizontal" (Sloterdijk, *Not Saved* 27). What happens in the experiential, empirical or ontical domain – again, for Stiegler, the realm of technical exteriorization – is but the steady unfolding of what is essentially determined from the outset at the inception and nothing within this domain can fundamentally change the trajectory. Only "the vertical blow of the immense" [*Vertikaleinschlag des Ungeheuren*] (Sloterdijk, *Not Saved* 22), meaning a turning of/in Being in the sense of "an upheaval of the total sense of Being" as it induces "an integral turn in the drift" or a "change in direction in the pull of [...] Dasein as a whole" (35) can overcome the danger of technology and inaugurate something entirely new.

As Michel Haar also observes quite perceptively regarding Heidegger's view of the history of Being:

> What controls all History [...] is the exigency of the inaugural, the *Anfang*: the first *Gewesenes*, the first essential-having-been, the first grasp of being – such is the force of the Greek *aletheia* [...] History is not only what is already there, but it is what is already accomplished in the destiny of the Commencement [i.e., the Greek inception; P.L.] [...] *Geschick*, the destiny of Being, "contains" in advance the totality of History's possibilities [...] The commencement persists throughout the entire History of Being... [abiding] unalterable, unchanged in all the sequences arising from it. (69–70)

Sloterdijk for his part talks about this implacable holding power of the inceptual *aletheia* as "the fuse of truth" which runs irrevocably from Ionia to Los Alamos (*Not Saved* 186), from the Artemision to the atom bomb we could say.

For Sloterdijk, Heidegger shows himself here to be a heir of Plato and Augustine in that he perceives human existence and becoming in the empirical sense as a movement of falling, of being caught in a false, inauthentic movement, of a *perversion* of the truth, whilst the salvation out of this situation consists of a "gathering back in that which is authentic" (Sloterdijk, *Not Saved* 34) understood in terms of what Plato called *metanoia* or *periagógé*, a reversal of the soul (35) or what Augustine referred to as a *conversion* toward the truth, i.e., toward the true Christian religion (39). And this reversal is thought of as something that can only be induced vertically: Heidegger's turning into the event of appropriation as a reversal of the soul (to wit: from the *homo animalis* to the *homo humanus*) is therefore much closer to Augustine than to Plato – much more "catholic" as Sloterdijk phrases it – in that he also believes humans are incapable by themselves of accomplishing this conversion, since it can only be "granted" and "effectuated" by God (Augustine) resp. Being (Heidegger). Heidegger's talk about the "favor" that resides in the danger – "favor" being a word that for Heidegger translates Heraclitus' notion of *philia* (*Heraclitus* 98) – is associated by Sloterdijk with the Christian term "grace," which he sees returning in Heidegger in his discourse on releasement (*Not Saved* 39–40). And just as Augustine announced the true, i.e., Christian religion through the redemptive conversion of the fallen human being – the heathen of Antiquity – to the new Christian "God-Man" (37–38), Heidegger for his part announces according to Sloterdijk a more profound, ontological–aletheiological "religion of the clearing" (Sloterdijk and Heinrichs 113) in which the human is to be transfigured, "elevated," from the fallen, forgetful *animal rationale* to Da-sein – to being the attentive and caring "Da" of the clearing. For Heidegger this clearing, the "clearing of being" [*Lichtung des Seins*], is thought of as an *original* openness that is prior to and independent of any technology, and that is only more and more occluded and forgotten with the progression of technology, or what Stiegler calls technical exteriorization.

## the "horizontality" of technology as exosomatization in stiegler

Stiegler precisely criticizes Heidegger's neglect of the "horizontal" in Sloterdijk's sense in what he calls his "transcendental discourse" on Being and Dasein (*Disorientation* 5) and accuses him of completely neglecting, like all metaphysics before him, the horizontal process of technical exteriorization or exosomatization that is *actually* behind the opening of the clearing and sustains this openness, although it can also occlude or even close it off, given its pharmacological nature.[8] What most originally characterizes Dasein for Stiegler is its "exosomatic condition" and that is to say its *technical* condition, rather than being endowed – through an allegedly irreducible, and frankly miraculous, moment of revelation (the Greek inception) – with an originally pristine and full-blown openness granting it authentic access to the ontological difference. It is its exosomatic condition that is irreducible, i.e., the fact of its being constituted and conditioned by technical artefacts, and the history of being for Stiegler is therefore first of all the history of exosomatization (*Neganthropocene* 249) – technical artefacts in their constantly changing interrelationships forming the "exosomatic soil" of the history of being (*Nanjing* 293). And any "clarification of the clearing" (Sloterdijk, *Not Saved* 89), any understanding and questioning of being by Dasein, as Stiegler lays out most elaborately in *What Makes Life Worth Living*, is always preceded by a technological – i.e., pharmacological – mutation or upheaval, resulting from the process of exosomatization, an upheaval that first of all puts Dasein itself in question before *it* can question the Being of beings (107–09).

Criticizing Heidegger with Marx and Engels, who were the first (in *The German Ideology*) to perceive the importance of the process of exosomatization, Stiegler in fact reproaches Heidegger for adhering to an essentially "idealistic" understanding of technics, situating its origin in "spirit" – a term frequently used by Heidegger in relation to the essence of technology and also used by Hui as we have seen above – instead of recognizing, as Marx and Engels did, that "spirit" originates from technics and that is to say from the process of exosomatization, thought by the latter in terms of transformations in the means of production (Stiegler, *Neganthropocene* 246). Stiegler thus proposes a materialist reading of the whole Heideggerian edifice and insists that the clearing or unconcealment exists only under the condition of exosomatization, from which Dasein cannot be cured since this would entail its disappearance as the "placeholder" of the ontological difference, i.e., the closure of its ontological openness and freedom (249). For Stiegler as for Sloterdijk the ontological difference, which Heidegger thinks transcendentally ("spiritually"?) and purely "evental" [*ereignishaft*] in complete independence of all causality and all "effectiveness" [*Wirksamkeit*], is the result of *work*, of invention, of technical production: "Truth, *alētheia*, is set into work [*mise en oeuvre*], is *Werk*, that is, work, machining, fabrication" (264).[9] And this, of course, is anathema to Heidegger, who time and again insists, and abundantly so in the *Black Notebooks*, that "beyng could never be explained on the basis of beings" and that it is indeed "[u]nexplainable and ineffective" (*Ponderings VII–IX* 227) or that "Being as transcending all beings can never be founded on it and be explained from it" (*Vigiliae* 114). And inceptions, also the other inception that is the turning into the event of appropriation, "never need effectivity" (*Ponderings VII–IX* 296). They simply "*are*; without effecting, their being compels a beginning again" (296).

Accordingly, as Stiegler emphasizes, Heidegger says nothing about the relation between the Greek inception and the adoption by the Greeks of alphabetic writing technology, which had revolutionary noetic effects on the originally oral Greek psychic and collective life. These became thereby literalized, opening up the very possibility of philosophy as the questioning of "the being of beings" and thus inaugurating the history of the West as the history of Being sensu Heidegger, becoming the history of metaphysics and culminating in our days in the reign of enframing. This literalization – which Stiegler also refers to as literal "*grammatization*" or the "*becoming-letter* of sound and word" (*Symbolic Misery* 54) – is the crucial case of technical exteriorization that allowed for the birth of the scientific and technological culture of the West – and more generally of the Western process of "*psychic and collective individuation*" (50). It is this (mnemo)technological rupture that caused the "Greek miracle" which Heidegger evokes so brilliantly in all its splendor yet forgetful of its technological condition of possibility in the *Introduction to Metaphysics*, as the "great inception" and that is to say as the *decisive* "inceptive opening up of the essence of Being-human," understood as the gathering and apprehending of the Being of beings (186).

This "Being-human" – as Dasein – is for Stiegler *decisively* grounded in the *technical* facticity of being "*within literality*" (*Nanjing* 142). And though this initially defines only Greek "Being-human," it later on becomes – via the Roman alphabet – the basis of Western "Being-human." Still later, with the process of colonization and closely accompanied by what Derrida has called "globalatinization" (Stiegler, *Symbolic Misery* 55), it attains global dominance through the imposition of this "*intellectual technology*" on non-Western psyches or "*spirits.*" These psyches are thereby alphabetically "grammatized" and as such alienated (57) and controlled "*through the control of their symbols*" (55), insofar as these are supported by this intellectual technology that is the alphabet, which has given rise in our time to digital network technologies, which for Stiegler represent the concretion of enframing (and which Heidegger designated as "cybernetics"). As he explains this process to Chinese students in one of his recently published *Nanjing Lectures*:

> you are Chinese, and not Greek. And yet, throughout this process of grammatization that unfolded in Greece, then in Western Europe, passing through the *mathesis universalis*, which for Leibniz was partly inspired by Chinese writing, then in capitalism, which is, today, advancing more rapidly in China than anywhere else – throughout all of this, Greece has become *our* history, in *you*, who are Chinese, as in *myself*, a Frenchman. (142)

## hui's allegiance to heidegger. reviving the *spirit* of cosmotechnics

So the Greek inception rules supreme, on a global scale now, but not because the Chinese and other nations are "urged" by Westerners in some inexplicable way – that is to say within the Heideggerian perspective – "into the realm of hearing of that claim of Being which speaks from the innermost core of modern technology," as Heidegger could have argued (*On Time* 7). Or is this claim – the imperative of enframing – precisely inculcated through the imposition of Western intellectual technology, thereby overruling the Chinese cultural heritage which Hui, re-interpreting it in terms of a cosmotechnics, aims to "resurrect" and renew? And are the Chinese, as well as other nations, as a result of this imposition "entrapped" just as Westerners in the *danger* of being closed off from their own "originary" clearing? Probably so, and this would amount to an annexation of Chinese culture through the imposition of an "intellectual technology" originating from a Western process of exosomatization, or as Hui puts it: through an enforced "*adaptation* to the technological condition" of the West, seemingly rendering any return to a "proper," "authentic" origin impossible (*Cosmotechnics* 152). Now what I would like to ask here, and this will conclude my modest attempt to understand Hui's project of cosmotechnics as wavering between Heidegger's and Stiegler's view of the relation between technology, ontology and the world (or the cosmos), but also developing something new of course, is how Hui views the relation between technology and ontology, and how he conceives of the local cosmotechnical "bifurcations" or "fragmentations" that he imagines not only for China but in their own manner for other non-Western cultures as well, as against the global "bifurcation" of the "technosphere" proposed by Stiegler as well as the singular "turning" of enframing envisaged by the late Heidegger.

Although he also conceives of cosmotechnics "as the possibility of unconcealment," more or less identifying the cosmos with the unconcealment and stating that "unconcealment happens in technical activities" (Hui, *Cosmotechnics* 16) – thereby supposing, I assume, that it emerges as a result of technical exteriorization – what I find remarkable is that every time he clarifies what he actually means with his concept of cosmotechnics, Hui uses notions such as "thinking" or "thought" (*Cosmotechnics* 7, 33, 43, 65, 101, 191, 197, 201, 217, 309; "Cosmopolitics" 8, 9; "Renewed Relation" 8, 14–16, 20; *Recursivity* 25, 39, 263, 269, 274, 276–78; "Cosmotechnical Event" 18), "spirit" (*Cosmotechnics* 79, 85, 164, 168; *Recursivity* 31, 232), "culture" (*Cosmotechnics* 6, 10, 14, 19, 217–18, 221, 242; "Cosmopolitics" 7–9; "Renewed Relation" 3, 17, 19; *Recursivity* 27, 30–31, 39, 223, 272; "Cosmotechnical Event" 15), "episteme" and "epistemology" (*Cosmotechnics* 31, 280, 296, 301, 307, 308, 310; "Cosmopolitics" 9; "Renewed Relation" 17; *Recursivity* 223, 226, 263, 265, 268, 275–78), "metaphysics" (*Cosmotechnics* 9, 29, 30–31, 34, 218, 296, 307), "form of life" (*Cosmotechnics* 31, 309; "Cosmotechnical Event" 16), "meaning" (*Cosmotechnics* 217) and recently also "sensitivity," "sensibility" and the "sense of existence" ("Renewed Relation" 18–19; *Recursivity* 263, 267–68, 276), sometimes also "ontology" (*Cosmotechnics* 280).

All these notions are terms for what we could call in the most general way *noesis* or the "noetic," i.e., thinking or knowing. This means that cosmotechnics for Hui is definitely related to thought and knowledge, and to a sensibility or sense of living, and apparently insofar as these cannot be fully reduced to the process of technical exteriorization and the effects thereof, or his antinomy or rather distinction between the universal dimension of technical exteriorization and the particular dimension of cosmotechnics would make not much sense. Indeed, this thinking and sensibility is specified as cosmological, but also as philosophical and sometimes mythological (Hui, *Cosmotechnics* 10–12, 14–17, 29) and we are assured that this thinking of the cosmos is itself always conditioned by technics, but when he concretely analyzes Chinese cosmotechnics he does so through an interpretation of fundamental *concepts* such as *Dao* and *Qi*, similar to how Heidegger re-interprets notions such as *physis*, *logos*, *techne* and *idea* as fundamental concepts of the Greek inception, a crucial difference being that Heidegger's focus is ontological whilst that of Hui is cosmological.[10] And while Hui considers cosmology to be the "larger configuration" from which the ontological should be interrogated (10), for Heidegger "cosmos" [Greek: *kósmos*] is just one "originary determination of being" (*Heraclitus* 134), next to others such as *physis*, *logos* and *aletheia*, referring to "world" or "worlding" as "originary adornment" in the sense of "the jointure of the conjoining of beings" (124). And whilst for Heidegger it is Being that represents the ultimate ground and genetic driver of technology, for Hui it is a particular cosmology (*Recursivity* 223) and an always local "cosmic reality" (39) that both enables and constrains the process of technical exteriorization, the latter being unthought in Heidegger.

The idea that *kósmos* in this ontological sense of world or worlding and cosmology as an *understanding* of it might be the product of technics is unthinkable for Heidegger, while it forms one of the presuppositions of Hui's idea of cosmotechnics. The notion of cosmotechnics seems to suggest more of a reciprocal conditioning of cosmology and technics, also in the sense of a co-evolution. But when he talks about cosmotechnics as an "*Urtechnik*" (Hui, *Recursivity* 271) – reminiscent of Heidegger's notion of *techne* but then one that is different depending on its cultural origin – Hui seems to have, not unlike Heidegger, something "spiritual" in mind, despite the fact that he is fully aware that the organological approach to technics that he shares with Stiegler precludes opposing technics and spirit. Yet, "[t]echnology is a product of the spirit," Hui explicitly asserts (*Recursivity* 31) and what cosmotechnics professes is a unification of technological instrumentality "with the spirit" (232). And it is this *spirit* that is in danger of being overwhelmed by Western technology, in danger because increasingly obstructed to "exercise its freedom" (30), not unlike the disempowered spirit of Europe notoriously evoked in 1935 by Heidegger in the *Introduction to Metaphysics* we may suggest (47), though from an entirely different point of departure and with an entirely different concern in mind.

This "exercise of freedom" by the Chinese "spirit" may be related to Heidegger's idea of acquiring a "free relationship" to technology (*Technology* 3), which would be the diametrical opposite or so it seems of the "becoming 'free'" for technology of the Chinese people under communism, which Heidegger predicted in one of his *Black Notebooks* from the 1940s (*Anmerkungen* 441) and which Hui interprets as their surrender to enframing and their falling prey to the same blindness and obedience to technology as Westerners (*Cosmotechnics* 6). This free relationship for Heidegger entailed a renewed experience of the openness of Being and a "retrieval" by Dasein of its ontological freedom, thereby also inaugurating a new "question-worthiness" [*Fragwürdigkeit*] of Being as the origin and inception of their technological modus vivendi. Coming from a different origin though, the Chinese cannot and should not await or prepare for a turning sensu Heidegger but instead re-appropriate and re-imagine their own cultural–cosmological sources – *re-thought as cosmotechnics* now that the organological condition has become irrefutable and inescapable – so as to transform their implication in enframing and bifurcate or fragment toward a different future in their own way. And the same goes for all other non-Western cultures, such as that of the Amerindian, the African and the Australian indigenous peoples, but for those in an altogether different way to be sure, if only because of their history of intense colonization and their largely oral traditions.

I contend though, and this may be thought of as a kind of "Heideggerian correction" to the onto-anthropological and Western-centric understanding of technology developed by Stiegler and Sloterdijk, that the idea of "technodiversity" (Hui, *Recursivity* 27) that Hui puts forward in the sense of a multiple cosmotechnics is first of all a *noodiversity*, and not so much understood with Stiegler as a project of differential adoption of the global "digital condition" (although this forms part of it as well), but as a culture-specific plurality of "dormant" or forgotten overall noetic dispositions or general forms of thought (be they cosmological, metaphysical or mythological) *and that is to say* in the sense of an *ethnodiversity* to be reclaimed and revived as *local grounds* from which this differential adoption may occur, thereby opening the possibility of a plural fragmentation or fractalization of world-history (265), or may we say: a plurality of "events of appropriation" not necessarily held to "Being" (but to other inceptual figures such as *sunyata* (nothingness), *mana* or *atman/brahman*). The Earth's technosphere as well as its noosphere, Hui seems to be pointing out, is still also an ethnosphere, despite the alarming loss of ethnodiversity it has experienced already (Davis 3–4), and it is precisely *there* that he situates the *local* seeds of a more pluralist future technosphere and noosphere or rather a techno-noosphere.

It is this frankly "ethno-ontological" approach to technology, imagining different, culture-specific "essences" of technology *as* cosmotechnics – in a sense combining Heidegger's postulate of an ontological (in Hui's case: cosmological) a priori driving the empirical development of technics with Stiegler's insight into the organological condition – that allows him to bring the philosophy of technology in dialogue with contemporary anthropology's *ontological turn* (Holbraad and Pedersen). It allows for instance to add a cosmotechnical perspective to the quadruple "geography of being" (Skafish) proposed by Philippe Descola in his *Beyond Nature and Culture* and imagine different cosmotechnical dispositions possibly present within the naturalist, animist, totemist and analogist ontologies that the latter distinguishes in this book (122). Or to confront the anthropologies behind Stiegler and Sloterdijk as well as the existential ontology of Heidegger with the Amerindian-inspired "counter-anthropology" – which can be understood as an entirely new, other-oriented kind of philosophical anthropology – crafted by Eduardo Viveiros de Castro (*Relative Native*), in which a "human" enters the scene that is *totally* different – and most probably also technologically – from the *anthropos* that took the stage as *to deinotation* and later the *zoon logon echon* in ancient Greece and imposed its technological destiny on all the other peoples of the planet.

However that may be, what Hui's project of cosmotechnics as the imagining of multiple technological turnings proceeding from multiple *cosmo-ontological* returnings suggests, is that non-Western cultural spheres should inquire into their *own* cultural archives and so recuperate their own cosmotechnical resources, in order to effectively confront their *own* danger of being engulfed and anestheticized by enframing, instead of relying exclusively on the Western archive. They should "remember" and actively revitalize their own ontological, cosmological and mythological origins and reclaim what Viveiros de Castro has so boldly called their "ontological self-determination" (*Cannibal* 43) so as to regain a chance of "technological self-determination," however problematic such an expression may appear to the average Heideggerian. Then, possibly, they might find out, as Sloterdijk suggests at the very end of his provocative 1999 lecture on the "Rules for the Human Park," that "even the archival cellars" may once again "become the clearing" (*Not Saved* 216) or contribute to its *always local* re-opening.

## disclosure statement

No potential conflict of interest was reported by the author.

## notes

1 I capitalize when I mean Being in the Heideggerian sense of the being (verbal) of beings.

2 Restricting myself to this interpretation here doesn't mean of course that I do not recognize the many other innovative and fertile features of his project.

3 I won't go into the moral dimension of cosmotechnics, which is crucial for Hui, considering its being defined as "the unification between the cosmic order and the moral order through technical activities" (*Cosmotechnics* 19).

4 It is this so-called "high altitude" understanding of "Technology with a capital T" that is emphatically dismissed by today's empirically oriented philosophers of technology (Ihde; Verbeek; see for a critique of this dismissal: Lemmens, "Thinking Technology").

5 Cf.

> Technics – [...] – is the pursuit of life by means other than life. And this is also the opening of what Heidegger believed should still be called the "question of being" as the advent of Dasein, that is, of the "being who questions." (Stiegler, *Nanjing* 45)

6 Indeed, as presencing itself, Being is nothing actual yet forms the "concealed basic trait of the actuality of everything now actual" (Heidegger, *Insight* 58).

7 In the onto-historical treatises such as *Contributions to Philosophy* and *Mindfulness*, but especially in the *Black Notebooks*, Heidegger characterizes the other inception as the event of appropriation in terms of a "divine intervention" or in his own words as an "encounter [*Ent-gegnung*] of gods and humans [with the strife of earth and world]" (*Ponderings XII–XV* 16), writing that:

> If the god necessitates beyng, and if the human being as Da-sein disclosively grounds the truth of beyng, and if a world arises out

of the abyss and the earth opens itself to bearing – the hour of a beginning has then arrived. (46)

Similar passages can be found all over the place, for instance on pages 83 and 94 or also in *Ponderings VII–IX* 110, 143, 149, 237.

8 As Stiegler writes in the first volume of *Technics and Time*, technology in the concrete sense "will never have had in Heidegger's thinking any *dynamic specificity*. It will have done nothing but follow the logic of the temporal fall into the historial forgetting of being [...] It will never have had the least *properly* unconcealing quality" (244).

9 Like Stiegler, Sloterdijk maintains that every clearing is a technical clearing and indeed that the clearing, i.e., unconcealment or *aletheia*, "is not to be thought without its technical provenance" (*Not Saved* 142). Criticizing Heidegger and his insistence on the originary "purity" of the openness of Being, he states that technology stands at the origin of unconcealment and therefore that the "human being does not stand in the clearing with empty hands – not like some destitute alert shepherd with his flock, as Heidegger's pastoral metaphors suggest"; instead, "the human being disposes over stones and the successors of stones, over tools and weapons" (142). Like Stiegler, who claims that Dasein's access to the clearing, contrary to Heidegger's point of view, "is not alienated but *constituted* by technics, and made possible by technicity" (*Epimetheus* 49), initially by stone tools (159), Sloterdijk argues that the clearing is originally "a *work* of stones" (*Not Saved* 116; my emphasis), which implies "that it should be thought from below" and that is to say: ontically, although in view of its ontological "height" (106). This amounts to a "dignification" of the question concerning technology in an ontico-ontological sense not unlike what we find in Stiegler (Lemmens, "From Ontology"). In a way that is strongly reminiscent of Stiegler's technological and technogenetic reinterpretation of Heidegger's existential ontology and history of Being in the *Technics and Time* series, though inspired by German philosophical anthropology (mainly Gehlen but also Alsberg and Claessens) rather than the paleoanthropology of Leroi-Gourhan, Sloterdijk anthropologizes and historicizes the clearing and attempts to show that its emergence within the human being results from a long process of self-domestication through basically four kinds of evolutionary, anthropogenetic mechanisms – which are at the same time "aletheiogenetic" mechanisms since they produce "the human capacity for truth" in the process (*Not Saved* 104) – which are described in considerable detail in the essay "The Domestication of Being" in his 2001 book on Heidegger (*Not Saved* 89–148). It is through this *technogenetic* process of anthropogenesis that the human species is gradually drifting out of its "truth-less" and closed animalistic "environing world" [*Umwelt*] and attains the open and "truth-relevant" world in the Heideggerian sense of the clearing. In this sense the world, or what may also be called the cosmos here, comes into being for Sloterdijk through technics, although he lays less emphasis than Stiegler on the technical artefact as such and focuses more on the technically mediated mechanisms of domestication and immunization, for instance through "anthropotechnics." These also form a crucial element of the "radically historical" onto-anthropological theory of "sphero-poiesis" and "sphero-immunology" laid out in the *Spheres* trilogy, and it can be argued that the "spherology" and the "general immunology" developed therein resonate with Hui's idea of cosmotechnics (Sloterdijk, *Sphären*).

10 As Hui writes: "I have also aimed to open up the concept of technics as multi-cosmotechnics, consisting of different irreducible metaphysical categories" (*Cosmotechnics* 307), where I take "irreducible" to mean not just irreducible to each other (e.g., Chinese to Greek) but irreducible to technology as technical exteriorization as well.

## bibliography

Davis, Wade. *The Wayfinders. Why Ancient Wisdom Still Matters in the Modern World*. Toronto: Anansi, 2009. Print.

Descola, Philippe. *Beyond Nature and Culture*. Trans. Janet Lloyd. Chicago: U of Chicago P, 2013. Print.

Haar, Michel. *The Song of the Earth: Heidegger and the Grounds of the History of Being*. Trans. Reginald Lilly. Bloomington: Indiana UP, 1993. Print.

Haff, Peter. "Technology as a Geological Phenomenon: Implications for Human Well-Being." *Geological Society London, Special Publications* 395 (2013): 1–9. Print.

Heidegger, Martin. *Anmerkungen I–V (Schwarze Hefte 1942–1948). Gesamtausgabe*. Band 97. Frankfurt am Main: Klostermann, 2015. Print.

Heidegger, Martin. *Basic Questions of Philosophy: Selected "Problems" of "Logic."* Trans. Richard Rojcewicz and André Schuwer. Bloomington: Indiana UP, 1994. Print.

Heidegger, Martin. *Basic Writings*. Trans. David Farell Krell. New York: HarperCollins, 1993. Print.

Heidegger, Martin. *Being and Truth*. Trans. Gregory Fried and Richard Polt. Bloomington: Indiana UP, 2010. Print.

Heidegger, Martin. *Bremen and Freiburg Lectures: Insight into That Which Is and Basic Principles of Thinking*. Trans. Andrew J. Mitchell. Bloomington: Indiana UP, 2012. Print.

Heidegger, Martin. *Die Geschichte des Seyns. Gesamtausgabe*. Band 69. Frankfurt am Main: Klostermann, 1998. Print.

Heidegger, Martin. *Discourse on Thinking*. Trans. John M. Anderson and E. Hans Freund. New York: Harper, 1969. Print.

Heidegger, Martin. *Heraclitus. The Inception of Occidental Thinking. Logic: Heraclitus's Doctrine of the Logos*. Trans. Julia Goesser Assaiante and S. Montgomery Ewegen. London: Bloomsbury, 2018. Print.

Heidegger, Martin. *Introduction to Metaphysics*. Trans. Gregory Fried and Richard Polt. New Haven: Yale UP, 2000. Print.

Heidegger, Martin. *Mindfulness*. Trans. Parvis Emad and Thomas Kalary. London: Bloomsbury, 2016. Print.

Heidegger, Martin. *On the Way to Language*. Trans. Peter D. Hertz. New York: Harper, 1971. Print.

Heidegger, Martin. *On Time and Being*. Trans. Joan Stambough. Chicago: U of Chicago P, 2002. Print.

Heidegger, Martin. *Ponderings VII–IX. Black Notebooks 1938–1939*. Trans. Richard Rojcewicz. Bloomington: Indiana UP, 2017. Print.

Heidegger, Martin. *Ponderings XII–XV. Black Notebooks 1939–1941*. Trans. Richard Rojcewicz. Bloomington: Indiana UP, 2017. Print.

Heidegger, Martin. *The Question Concerning Technology and Other Essays*. Trans. William Lovitt. New York: Harper, 1977. Print.

Heidegger, Martin. *Vigiliae und Notturno (Schwarze Hefte 1952/53–1957). Gesamtausgabe*. Band 100. Frankfurt am Main: Klostermann, 2020. Print.

Holbraad, Martin, and Axel Pedersen. *The Ontological Turn. An Exposition*. Cambridge: Cambridge UP, 2017. Print.

Hui, Yuk. "Cosmotechnics as Cosmopolitics." *E-flux Journal* 86 (Nov. 2017). Web.

Hui, Yuk. "For a Cosmotechnical Event." *Foundations of Science*. Forthcoming.

Hui, Yuk. "On Cosmotechnics: For a Renewed Relation between Technology and Nature in the Anthropocene." *Techne. Research in Philosophy and Technology* 21.2 (2017): 1–23. Print.

Hui, Yuk. *The Question Concerning Technology in China. An Essay in Cosmotechnics*. Cambridge: Urbanomic, 2016. Print.

Hui, Yuk. *Recursivity and Contingency*. London: Rowman, 2019. Print.

Ihde, Don. *Technology and the Lifeworld. From Garden to Earth*. Bloomington: Indiana UP, 1990. Print.

Lemmens, Pieter. "From Ontology to Organology. Heidegger and Stiegler on the Danger and Ambiguity of Technology and Technical Media." *We Need to Talk about Heidegger. Essays Situating Martin Heidegger in Contemporary Media Studies*. Ed. Justin Michael Battin and German A. Duarte. Berlin: Peter Lang, 2018. 279–307. Print.

Lemmens, Pieter. "Thinking Technology Big Again. Reconsidering the Question of the Transcendental and 'Technology with a Capital T' in the Light of the Anthropocene." *Foundations of Science*. Forthcoming.

Skafish, Peter. "Anthropological Metaphysics/ Philosophical Resistance." Theorizing the Contemporary, *Cultural Anthropology* (13 Jan. 2014). Web. <https://culanth.org/fieldsights/464-anthropological-metaphysics-philosophical-resistance>.

Sloterdijk, Peter. *Not Saved. Essays After Heidegger*. Trans. Alexander Moore and Christopher Turner. Cambridge: Polity, 2017. Print.

Sloterdijk, Peter. *Sphären*. Frankfurt am Main: Suhrkamp, 1998–2004. Print.

Sloterdijk, Peter, and Hans-Jürgen Heinrichs. *Neither Sun Nor Death*. Trans. Steve Corcoran. Cambridge, MA: Semiotext(e), 2011. Print.

Stiegler, Bernard. *Automatic Society Vol. 1. The Future of Work*. Trans. Daniel Ross. Cambridge: Polity, 2016. Print.

Stiegler, Bernard. *The Nanjing Lectures 2016–2019*. Trans. Daniel Ross. London: Open Humanities, 2020. Print.

Stiegler, Bernard. *The Neganthropocene*. Trans. Daniel Ross. London: Open Humanities, 2018. Print.

Stiegler, Bernard. *Symbolic Misery 1. The Hyperindustrial Epoch*. Trans. Barnaby Norman. Cambridge: Polity, 2014. Print.

Stiegler, B. *Technics and Time 1. The Fault of Epimetheus*. Trans. Richard Beardsworth and George Collins. Stanford: Stanford UP, 1998. Print.

Stiegler, Bernard. *Technics and Time 2. Disorientation*. Trans. Stephen Barker. Stanford: Stanford UP, 2009. Print.

Stiegler, Bernard. *What Makes Life Worth Living. On Pharmacology*. Trans. Daniel Ross. Cambridge: Polity, 2013. Print.

Verbeek, Peter-Paul. *What Things Do. Philosophical Reflections on Technology, Agency and Design*. Trans. Robert P. Crease. University Park: Pennsylvania State UP, 2005. Print.

Viveiros de Castro, Eduardo. *Cannibal Metaphysics*. Trans. Peter Skafish. Minneapolis: Univocal, 2014. Print.

Viveiros de Castro, Eduardo. *The Relative Native. Essays on Indigenous Conceptual Worlds*. Chicago: Hau, 2015. Print.

marco pavanini

# COSMOTECHNICS FROM AN ANTHROPO-TECHNOLOGICAL PERSPECTIVE

## I introduction

Yuk Hui's (*Question*) cosmotechnical approach represents a twofold attempt. On the one hand, it aims to acknowledge human constitutive technicity and hence understands technics as a universal anthropological category, in so far as every culture adopts technical practices and is structured by them in its psychophysical as well as eco-symbolic organisation. On the other, it stresses the differences among the technical activities each culture develops as well as among the different understandings of technics each culture elaborates.

Reconciling the potential for tension between these two tendencies is one of the foremost concerns of Hui's inquiry. In this paper, I will first underline some of the key hermeneutical contributions of the concept of cosmotechnics, concerning our understanding of the discomforts related to modern technologies and the process of global modernisation, before further analysing the relationship between the way a culture develops technics and its cosmotechnical understanding of the latter. By doing so, I aim to stress how the unification of the moral and the cosmic provided by cosmotechnics is originary, that is, how it establishes our understanding of the world as well as the possible variations within this understanding.

This inquiry relies upon the assumption that technics is not only what structures the human ontological constitution as well as its spatially and temporally bounded inner differentiation, but also what enables and defines every possible discourse and practice revolving around humans. This recursion should make us aware that the question of technics is constitutively not free from the operations performed by technics itself, as it is always deeply influenced by the techniques of the culture within which it emerges.

The standard definition Hui (*Question*) provides for cosmotechnics is "the unification between the cosmic order and the moral order through technical activities" (19). It is worth analysing this definition, as it singles out three different elements: the cosmic order, the moral order and technical activities, the last of these performing the unification of the first two. It would be inaccurate to claim that cosmotechnics represents how technics enables humans to unify their cultural behaviours with their understanding of the outer, "natural" environment through the usage of tools. We may have a better comprehension of Hui's view if we stress the constitutive and originary character of technical activity, which is not to conjoin elements that would otherwise have existed separately. The unification is inaugural: technical activity enables human cultures to have both a cosmic and a moral order, and to keep them in constant relation. Cosmotechnics provides inner coherence within a culture's worldview, establishing its way of relating to beings. What is understood under the concept of cosmotechnics is that humans organise their experience through technical practices, the latter not only defining habits, rites and value systems, but also concretely installing them into the surrounding worldly conditions. Each culture has its cosmotechnics, which defines it as *that* culture, differentiating it from the others: "What kind of morality, which and whose cosmos, and how to unite them vary from one culture to another according to different dynamics" (Hui, "Cosmotechnics as Cosmopolitics"). In a sense, cosmotechnics is something all-encompassing for the determination of a culture: "Human activities, which are always accompanied by technical objects such as tools, are in this sense always cosmotechnical" (Hui, *Question* 4).

Cosmotechnics's theoretical stance and contribution to philosophy of technology is thus twofold. On the one hand, it aims to overcome the metaphysical dichotomies which emerged with higher cultures and are especially widespread in the West. On the other, it strives to undermine the current conceptions of modern technology in relation to both its underestimation as mere instrumentality and its alleged universal significance.

## 2 cosmotechnics as a way to overcome the nature–culture divide

Habitual, Western philosophical and theological thinking as well as contemporary, scientific-led common sense tend to conceive reality as structurally split into two fields: the sphere of the spiritual, that is, the immaterial, intentional and spontaneous capacity of establishing concepts and norms, and the sphere of the material, namely, passive, corruptible and dependent physicality, always in need to receive form and direction (Esposito 16–98).[1]

Indeed, a cosmotechnical approach enables us to get rid of this dichotomy, since cosmotechnics is what makes explicit the interconnection between the moral, that is, the alleged interior, spiritual realm and the cosmos, namely, the so to speak outside world, understood as an ordered system of events and things. Thanks to its cosmotechnics, a culture is able to give sense to its way of living in a given space and time. Yet this sense-giving performance is not much a posterior explanation and justification of the given, it is rather a structuring schema of experience that lies at the bottom of collective life, providing it with – and performing in it – its specific direction and aim.

Furthermore, it is important to evidence that this operation of connection and unification is not to be understood as a reconciliation between two poles, theory and praxis, that would have used to be separated and distinguished. A culture's cosmotechnics is originary, in the sense that it opens up the possibility of existence and "thinkability" of that culture, representing the linguistic and symbolic as well as instrumental and institutional background, starting from which something like the cosmic and the moral order can be identified. That is to say that the latter are constructed through cosmotechnics, and do not exist prior to a cosmotechnics supposed to subsequently appear in order to put them together in a coherent frame. They are rather to be considered as a possible, determinate outcome of a given cosmotechnics, so to speak as two sides of one and the same event, and exist only in their reciprocal relationship, having sense in so far as they are conceived as what is structured and informed by a cosmotechnics. Foucault's concept of *episteme* may be of some help here, since Hui also recalls it himself in order to better explain his understanding of cosmotechnics:

> The different "cosmotechnics" can be further analyzed according to their cultural

> specificities and understood in terms of different or alternative epistemologies, as well as *episteme* in the sense of Michel Foucault [...] namely the relations between different scientific domains which define the regime of truth. (Hui, "On Cosmotechnics" 17)

According to Foucault, an *episteme* is what denotes the dominant system of knowledge in a given region and epoch:

> [W]hat I am attempting to bring to light is the epistemological field, the *episteme* in which knowledge, envisaged apart from all criteria having reference to its rational value or to its objective forms, grounds its positivity and thereby manifests a history which is not that of its growing perfection, but rather that of its conditions of possibility; in this account, what should appear are those configurations within the *space* of knowledge which have given rise to the diverse forms of empirical science. (xxiii–xxiv)

Hence, the *episteme* is not limited to structuring the modes of knowing and understanding belonging to scientific, philosophical and theological paradigms, and to the way people give sense to their world accordingly. It also defines a culture's practices and framing possibilities, directly influencing its institutions, habits and social structures. In this sense, the *episteme* represents a complex of discourses, i.e., a practical–cognitive schema structuring collective experience and understanding. It is indeed an apparatus (Agamben 1–24), i.e., a set of technical procedures producing a discrimination within the continuum of embodied experience, enabling or disabling given performances of the ones undergoing it.

This unification of the moral and the cosmos performed by technical media becomes perhaps more evident if we look at the phenomenon of consumption of psychotropic substances some cultures perform in order to improve their social synthesis (Sloterdijk, "Wozu Drogen?" 118–60). In these cases, a technical apparatus, that is, drug consumption, is collectively endorsed to produce a sense of co-belonging through the communication with a transcendent sphere. Here the moral order, that is, the complex of mutual duties and obligations a society is made up of, is connected to the cosmic order, that is, the deities embodying "natural" forces and entities. However, the consumption of the substance is meaningful and profitable for the culture only if performed in a strictly ritualised environment, i.e., as long as it is established in its dosage, timing and context. Hence, such cultures' cosmotechnical apparatuses are in turn the core of their system of decisions concerning how to schematise and categorise their experience of a common life in a given *milieu*.

As we have seen concerning the theory–praxis combination, another way of manifestation of this old dichotomy is represented by the distinction, if not opposition, of nature and culture, in so far as the former is identified with what characterises humanity, and the latter with what humans share with the rest of beings. Western modernity largely relies upon this binary distinction (Sloterdijk, "Time of the Crime" 281–95), in order to promote a worldview set on two separate dimensions, the former based on the spiritual, that is, the soul or the mind, the latter based on the material, that is, the body or the physical. As argued by Descola (*Beyond Nature*), the former is generally understood as what all human cultures share and belong to, whereas the latter represents what differentiates them from one another (172–200).

Once again, with cosmotechnics we can state the only apparent character of this dichotomy. Cosmotechnics being originary, it represents the plan from which a culture's practical–cognitive schemas can develop, enabling it to adopt one classification of being rather than another. So, what is understood as the cosmos, namely, the allegedly outer part of reality, is actually deeply cultural, since it is from bottom up framed and structured through symbolic grids and institutionalised habits. At the same time, what is conceived as the moral, that is, the alleged cultural realm, is intrinsically natural, being structurally influenced and determined by biological, ecological and environmental conditions, in turn provoked by and structured through tools- and language-mediated practices enabling the emergence of its complex of obligations and prohibitions.

If we follow Latour's (91–129) insights on this point, as also suggested by Hui, some light can be shed on this matter. Cosmotechnics is indeed the technically mediated epistemological dimension, starting from which a culture is able to perform its schematisation of experience. It has a phenomenotechnical significance, in the sense in which Latour and Woolgar reinterpret Bachelard's insights on scientific laboratory activity:

> It is not simply that phenomena depend on certain material instrumentation; rather, the phenomena are thoroughly constituted by the material setting of the laboratory. The artificial reality, which participants describe in terms of an objective entity, has in fact been constructed by the use of inscription devices. Such a reality, which Bachelard [...] terms the "phenomenotechnique," takes on the appearance of a phenomenon by virtue of its construction through material techniques. (Latour and Woolgar 64)

We can summarise by maintaining that the construction of scientific, philosophical and political performances is always framed by a determinable complex of tools and symbolic apparatuses and should therefore be understood without prescinding from it. Only starting from this epistemic position, can something be labelled as nature or culture. According to Latour, this distinction has a performative value for the culture's rules of living, but it is not to be forgotten that it emerges only subsequently and dependent on the cosmotechnical practices structuring it. Nature and culture are not only temporally and spatially situated concepts; they also exist only within their reciprocal opposition and dependence.

## 3 human originary technicity and cosmotechnical plurality

The concept of cosmotechnics also bears great explanatory potential concerning the contemporary understanding of technics. Following Stiegler's (*Technics and Time 1*) position, Hui claims that technics' pivotal importance has been neglected within Western thought, representing what unconsciously underlies its understanding of the human. Hence, technics has traditionally been conceived as mere instrumentality, i.e., as something that is just added to the ontological constitution of the human, without being a structuring part of it.

To conceive technics as instrumentality means to underestimate the technical nature of humans, amounting to an understanding of technical artefacts as things that are just used by them, without bearing any feedback on the way they frame their worldview and behaviour. Technics would thus be an accidental supplementation to an already complete living being able to dispose of it according to its will. Positing an originary independence between technics and the human elicits, only apparently paradoxically, the sense of impotence and submission people experience when their technical systems evolve to the extent of encompassing and reframing their whole life conditions (Moore 68–75). After the industrial revolution and up to capitalism's contemporary stage, which actually represents "*the* contemporary cosmotechnics that dominates the planet" (Hui, *Question* 299), humans, firstly in the Western world and then in most countries over the globe, undergo a passive submission to their own machines. Since the latter are considered as not influential for the framing of experience of the humans, now that they appear to be so dominant and pervasive, they are perceived as prevaricating and hegemonising humans' possibilities of subjectivation. Hui refers here to the point made by Simondon, in which he claims that this false understanding of technology elicits contemporary technophobic behaviours and the related discomforts they both provoke and are triggered by, especially concerning the feeling of being overwhelmed and superseded by machinery (103–63).

Since technics and humans are originally co-belonging, in the sense that the former produces the latter to the same extent as the latter produces the former (Sloterdijk, "Domestication" 89–148), discomforts concerning the understanding of the contemporary technical system and the effects it provokes, i.e., its disruptive perturbation of the previous life conditions, have to be understood as modifications of this constitutive relationship being at least as old as the genus *Homo*, and handled accordingly. As stated already by Leroi-Gourhan and confirmed by Hui, humans are technical beings through and through. They trigger the evolution of their technical system together with its potentially disruptive becoming and therefore the solution to the problems this system may elicit are to be found in a different cosmotechnics, i.e., in a new constellation of the relationship between the humans, their moral orders and habits, and the cosmos, the complex of events and things within which humans exist, framed by their technical artefacts and procedures as well as by their symbolically mediated understandings.

Starting from what I outlined so far, we are now able to approach Hui's interpretation of Heidegger's famous essay *The Question Concerning Technology* (3–35). Heidegger was one of the first explicitly posing the issue of technics and emphasising its relevance in order to grasp the specificity of our epoch. Nonetheless, according to Hui (*Question*) the centrality of the question of technics for the understanding of the humans has to be implemented within a cosmotechnical perspective,

> [f]or it implies a tacit acceptance that there is only one kind of technics and technology, in the sense that the latter are deemed to be anthropologically universal, that they have the same functions across cultures, and hence must be explained in the same terms. (4)

Interpreting Heidegger's (3–35) stance and in so doing turning it upside down, his claim that technics "is a mode of revealing" (13) acquires a new meaning in Hui's theoretical frame, namely, that humans shape their conception of the world and define the way they inhabit it according to the peculiarity of the technologies they adopt and invent, while for Heidegger it is the historial configuration of the originary human disclosure of Being that enables them to perform technics in one way or another.

According to Heidegger, this understanding of technics, and of modern technology as *Gestell*, i.e., as what exploits an allegedly endlessly available supply, despite coming from Western philosophical tradition, is also the only possible one, not only due to its capacity to indefinitely impose itself and expand, but also and above all because technics is considered as a matter uniquely pertaining to the West. Shifting the theoretical focus from technics in general to different cosmotechnics amounts to considering that the understanding of technics is not something immediately universal, but rather a phenomenon strictly depending on a culture's specificity. Every culture, i.e., every temporally and spatially bounded human group provided with inner coherence concerning its habits and institutions, bears an original understanding of technics. This understanding, in turn, determines a culture's specificity and its possibility to cope with the becoming-global of the contemporary technical system: "Technics is not anthropologically universal; technologies in different cultures are affected by the cosmological understanding of these cultures, and have autonomy only within a certain cosmological setting – technics is always *cosmo*technics" (Hui, *Question* 19).

The necessity of inventing new cosmotechnics, preserving their plurality and partial autonomy, comes from the urge to face up against the general tendency of modernisation, amounting to the assimilation of every culture to the same, Western, i.e., capitalist cosmotechnics. Furthermore, this cosmotechnics, being not explicitly thematised as such to date, hides the constitutive relationship between humans and technics, acting as a mere imposition of contemporary capitalism's market with its disruptive operations, regardless of the specificity of each culture, while according to Hui "the central idea is that every non-European culture must systematize its own cosmotechnics and the history of such a cosmotechnics" ("Cosmotechnics as Cosmopolitics").

## 4 cosmotechnics and the ontological turn in anthropology

These assertions lead us to a very interesting aspect of the theory of cosmotechnics. On the one hand, Hui acknowledges the constitutive technicity of the human and states that technics is a universal phenomenon, since humans become humans and endure, produce and transmit their humanity only thanks to their relationship with technical artefacts, "if we understand the concept [of technics] to denote skills for making artificial products" (*Question* 8). Hence, "*technics* refers to the general category of all forms of making and practice" (4 n. 3) and must be distinguished from cosmotechnics, that is, the understanding of technics provided by each culture, as well as from *téchne*, i.e., ancient Greek cosmotechnics, and technology, denoting the contemporary, Western technical system. As we have seen above, Hui here follows Stiegler and Leroi-Gourhan, claiming that the technical exteriorisation of knowledge and knowhow has to be conceived as a universal phenomenon defining human nature. Yet this assumption, left alone, would leave unquestioned the reason why in different cultures there are different technical systems, that is to say different artefacts as well as evolutions of these artefacts and of the way they interrelate:

> There is a general misconception that all technics are equal, that all skills and artificial products coming from all cultures can be reduced to one thing called "technology." And indeed, it is almost impossible to deny that technics can be understood as the extension of the body or the exteriorisation of memory. Yet they may not be *perceived* or reflected upon in the same way in different cultures. (Hui, *Question* 9)

Leroi-Gourhan tries to give an account of this differentiation, but only succeeds partially, in so far as he attributes this diversity to strictly environmental, that is to say "natural," nonhuman conditions (299–311). Each culture would develop its own technologies because they would fit in the pregiven life conditions this culture is confronted with. Hui (*Question*) implements this assumption, claiming that it is not merely environmental factors that contribute to the differentiation of a culture's technical *milieu*, but also its cultural environment, that is to say its worldview, social order and institutions:

> Yet even if we agree with Leroi-Gourhan in seeing the exteriorisation of memory as a general technical tendency, this does not yet allow us to explain why and how each culture exteriorises at a different pace and with a different direction; that is, it does not explain how exteriorisation is determined by certain conditions – not only biological and geographical, but also social, cultural and metaphysical. (217–18)

It is not a strict ecological determinism that is solely responsible for the variations within human technical environments, as Leroi-Gourhan assumes. Technics in general, i.e., the human tendency to exteriorisation in the inorganic, and cosmotechnics, in the sense of different, culturally related understandings of this exteriorisation, should be more precisely distinguished:

> [T]he experience of technics is related to and partially conditioned by cosmology – and it is precisely in this sense that we insist on the importance of a *cosmotechnics*. Technical apparatuses function somatically as extensions of organs – and, as prostheses, are *somatically and functionally* universal, and yet they are not necessarily *cosmologically* universal. That is to say, in so far as technics is both driven by and constrained by cosmological thinking, it acquires different meanings, beyond its somatic functionalities alone. (Hui, *Question* 217)

If the general, functional meaning of technics is universal, the account a culture gives to this phenomenon is quite variable and not limited to mere theoretical reflection. It is rather essentially cosmotechnical, i.e., it frames the culture's worldview, giving sense and coherence to its worldly condition and structuring its possibilities of experience. In order to better comprehend this, it may be of help to recall some assumptions from what has been characterised as the ontological turn in anthropology (Holbraad and Pedersen), i.e., a theoretical stance assumed by some prominent contemporary anthropologists such as Latour, Descola (*Beyond Nature*), Ingold and Viveiros de Castro.

The general aim of the ontological turn can be considered as analogous if not complementary to the cosmotechnical approach. Indeed, these anthropologists and social scientists defend the need to "take seriously" (Holbraad and Pedersen 155) the worldviews proposed by the cultures that form the objects of their ethnographic research. In this sense, their conceptions of the different beings and of the relationships subsisting among them are defined as ontologies, in order to stress the constitutive intention underlying these approaches (Descola, "Constructing Natures" 82–102). The Western overarching, ethnocentric mode of thinking, which tends to understand these worldviews as naive, if not distorted variations of an allegedly true conception of nature, has to be overcome, as it is argued that these accounts possess an equal pretension to truth and effectiveness as the one promulgated by the modern West.

Usually, modern ethnographic accounts consider cultures to represent different perspectives on a common nature proper to all humans and the criteria determining what this nature should consist of are developed from the Western *episteme*. This implies that each worldview is understood in a comparative way starting from the anthropologist's practical–cognitive schemas of experience, basically pretending to teach the other cultures what the nature they conceive differently should be in reality. Furthermore, the more these worldviews differ from the Western one, the more they are considered as eccentric, outdated or superstitious. They are submitted to this dominant, globalising *episteme* and considered as its distorted variants. On the contrary, the ontological turn claims that there is not a nature common to all cultures as well as there is no common culture. There are only different couplings of nature and culture, if we want to keep for a moment the traditional Western dichotomy (Descola, "Societies of Nature" 107–26). Each coupling, that is, each ontology, is equally legitimate and has to be understood, as much as possible, through the epistemic paradigms proper to the culture endorsing it. As stated by Descola (*Beyond Nature*),

> it is not so much linguistic limits, the perimeter of a commercial network, or even the homogeneity of modes of life that mark out the contours of a collective. Rather, it is a way of schematizing the experience shared by a more or less vast collection of individuals, a group that may well present internal variations – of languages, institutions, and practices – that are sufficiently marked for one to consider it, on a different scale, as a transformational group composed of separate units [...] If [...] one recognizes that the limits of any collective are coextensive with the area of influence of this or that schema of practices, then its definition will depend above all on the manner in which the humans in it organize their experience, in particular in their relations with nonhumans. (176)

The same can be stated about cosmotechnics: as there is no common understanding of the organised outer world, there is neither a common understanding of technics, although the latter is something shared by all humans, exactly as it is fair to claim that all humans frame their respective environments according to specific, determinable modes of classification. Although acknowledging the heuristic potential of this stance and recalling its value for his analysis, Hui reproaches the ontological turn for neglecting technics, i.e., for exclusively focusing on how cultures understand their relationship with nonhumans, without taking sufficiently into account how their ontologies are always structured through technical activities:

> [T]his tendency also suggests that the question of technics is not sufficiently addressed in the ontological-turn movement. For example, Descola talks often of practice, which may indicate his (laudable) desire to avoid an opposition between nature and technics; but by doing so, he also obscures the question of technology. ("Cosmotechnics as Cosmopolitics")

From Hui's viewpoint, Descola's ontologies should be understood as cosmotechnics: not only is it the case that each culture structures its own worldly experience with the same ontological dignity as the Western one does, but this experience is also framed through and allowed by the culture's particular understanding of technics, given that technical practices shape the human constitution and differentiate cultures from each other.

## 5 the problem of cosmotechnical comparativism

Under this viewpoint, we may notice that Hui seems somehow to waver between the perspective defended by the ontological turn and its opposite. As stated by Descola (*Beyond Nature*), it is the ontology belonging to a given collective that determines its possibility to adopt a specific technology, framing the becoming of its technical system in a particular direction and not in another: "It is not technical progress in itself that transforms the relations that humans maintain between themselves and the world but rather the sometimes tiny modifications made to those relations" (187).[2]

According to Hui, this stance is determined by the aforementioned neglecting of technics: "In speaking of a tension between ontology and technics I mean that these ontologies are only possible when they are already complicit with the technical life [...] As a result, any transformation of the latter will directly alter the former" ("On Cosmotechnics" 7–8). It is the technical becoming that a culture undergoes which determines the development of its ontology, that is to say its cosmology. The aim of a cosmotechnical perspective is to take into account how the transformations of the technical system influence the "unity of the cosmic order with the moral order," namely, to consider how these transformations enable the culture to acquire a worldview and prevent it from assuming another. However, elsewhere Hui may seem to lean toward the opposite conception. Taking the example of China, which serves as case study for its inquiry in cosmotechnical pluralism, he claims that "[t]he fact that Chinese culture does not elaborate on time and geometry, then, may have served as a cultural and cosmological condition of its technological development, producing, in Leroi-Gourhan's terms, different technical facts within the universal technical tendency" (Hui, *Question* 221).

Here the causal relationship between technological becoming and cosmology seems inverted, resulting into something quite similar to Descola's stance: it is the scarce consideration devoted in ancient China to the concepts of time and geometry that elicited the ontological presuppositions underpinning a determinate development of its technical system. A culture's understanding of technics, in so far as it frames its symbolic and institutional categories, would determine how and why a culture adopts certain technical artefacts, practices and instrumental systems.

From my viewpoint, both readings of Hui's theory would be excessively one-sided, either attributing only to a culture's technical becoming the explication of its worldview, or requiring from the latter alone to give account of its whole technical becoming. This apparent contradiction can easily be solved if we consider the existence of a circular recurrence between the two stances. This perspective is encouraged and implied by the concept of cosmotechnics itself, in so far as it fosters a "unification" between two orders, that as we have seen above are to be understood as co-belonging and interrelated from the very start. Thus, a culture's technical becoming frames its worldview, since different technologies enable different ways to structure experience, configuring the culture and its understanding of the world in a certain way. At the same time, a culture's cosmology also determines its conception of how the latter understands technics, channelling its technical becoming in a specific direction and preventing it from adopting other technological systems.

Considering the way different cosmotechnics may possibly interrelate, we may better appreciate how the same technology can be perceived in completely different ways by different cultures if we consider the aforementioned example of the ritualised consumption of psychotropic substances (Sloterdijk, "Wozu Drogen?" 118–60). If in some cultures the use of drugs is at the core of the production of the social synthesis and is therefore performed as what properly constitutes their sense of co-belonging, in

Western societies drug consumption mostly represents an escapist phenomenon that is symptomatic of people existing at the borders of the community, not being able or willing to effectively integrate. We have here basically the same technical practice, that is, the assumption of drugs, but with two radically differing functions. What determines this divergence is its mode of consumption, performing unity within the culture if experienced as what enables its components to have access to a common, transcendent sphere of meaning, or eliciting and worsening isolation and interpersonal disruption if perceived as a possibility to avoid a social environment with which it seems impossible to cope. Once again, the unification of the moral and the cosmic can be performed only via technical media, whose meaning may in turn sensibly differ from one culture to another, given that it is a culture's cosmotechnics that properly gives account of the function of the technical practices it develops.

For instance, within Australian Aboriginal communities the use of psychoactive substances, prescribed and managed within structured rituals, has been for a long time a common, shared means to achieve social cohesion and intergenerational bonding (Dobkin de Rios and Stachalek 155–61). Yet with the advent of colonisation by the West, the cultural background of these populations has been extensively disintegrated and their totemic cosmotechnics have fallen into pieces. This has elicited a shift in the use of some of their technologies, namely, drugs, this time mostly in the shape of Western imported alcohol, which is now consumed for a completely opposite goal, i.e., for the sake of a desperate retreat into a state of semi-perpetual hallucination as a response to the collapse of the former life conditions (Kahn et al. 351–66). Thus, while the technical practice remains quite the same, its use changes completely, because the native cosmotechnics supporting its adoption have been replaced with the disruptive agency of the Western cosmotechnical apparatus, suffered by the native populations as an unbearable lack of meaning in life.

Thus, without disregarding the constitutive technicity of humans and therefore the fact that every human culture performs feedback-bearing technical activities, a cosmotechnical thinking focuses our attention on the understanding these cultures may have of their own technical practices and hence of technics. Connecting cosmologies with their respective technical apparatuses enables us to conceptualise not only the strict relationship between worldviews and technical media, but also the different possibilities of understanding this relationship and therefore giving account of the world, attributing a meaning to it and enabling specific developmental possibilities.

Each culture has its own cosmotechnics or is forced to adopt the contemporary dominant, Western cosmotechnics of capitalism. Yet, as Hui asserts, a culture may or may not be conscious of its own cosmotechnical system. This means that it may have only an implicit understanding of technics and thus the way this understanding relates to its cosmic and moral order may not be thematised as such. Indeed, Hui's attempt strives toward the explication of these sometimes implicit cosmotechnical systems, as he broadly shows concerning China. Starting from this viewpoint, a question arises, concerning the epistemological position of the one performing the inquiry in cosmotechnics. Which cosmotechnics enables us not only to trace its own genealogy, bringing it to consciousness, but also to admit and understand the existence of a plurality of cosmotechnics, realising that there are other cosmotechnics not belonging to the culture to whom the one tracing this genealogy belongs? The same question could be asked in relation to the ontological turn in anthropology: if every culture frames its worldly experience and conceives the relationships existing among beings in its own way and if this way of categorising has its own reality and truth pretension, as we have seen, thanks to the technical media shaping its discourses and practices, then which ontology, which way of framing the experience enables us to thematise the existence of a plurality of ontologies and to acknowledge their ontological dignity?

However, this question is somehow ill-posed from the beginning: the problem is not which cosmotechnics or ontology allows this operation, since this abstractive and relativising performance is, at least potentially, proper to every culture, as we will see below, thanks to cultural techniques. The question would rather be: how and to what extent the cosmotechnics of the culture to whom the one performing the inquiry belongs influences and determines the result of the inquiry itself? A remark from Viveiros de Castro to Descola's approach seems to go in the same direction: "[i]n effect, its design makes it impossible for Descola's system to not predominantly express one of the four ontologies he identifies" (83).

Detecting its own cosmotechnics is a goal each culture can attain, at least in theory, and the result of this investigation would depend only on that culture's understanding of technics and use of technologies. However, the attempt to detect another culture's cosmotechnics would give different results according to the culture to whom the one performing this operation belongs, and this means according to its cosmotechnics. What I aim to show here is that each cosmotechnics constitutively influences the process of detection of other cosmotechnics. This is because cosmotechnics represents the complex of technical practices shaping a culture's worldview and structuring its worldly experience. Furthermore, it also frames its way of categorising knowledge and understanding itself and other cultures. For instance, Chinese cosmotechnics would not be understood in the same way if inquired starting from the implicit cosmotechnical apparatus of the Chinese only, or from the cosmotechnical standpoint of Western civilisation, or from some other cosmotechnical viewpoint. I would tend to claim that a culture's technical apparatuses not only influence its mode of structuring knowledge, but also condition the results of this knowledge, and this becomes relevant when the object of such knowledge is assumed to be something not belonging to the same *episteme*, that is, to the same cosmotechnics.

If we inquire into another culture's understanding of technics, paying attention to neither assimilate it with our understanding nor consider it as its primitive or deteriorated form, we should also be aware that the result of our inquiry would in any case be influenced by our own cognitive schemas of analysis. And these schemas are in turn determined by our cosmotechnics, that is to say by how our understanding of technics influences our symbolic framing of experience and worldview. We should consider that what matters here is to be properly aware of the operative influence our cognitive schemas bear on the result of the analysis, especially when, as it is the case, the object of inquiry corresponds to what determines a culture's cognitive structures, at the same time being recursively influenced by it.

## 6 cultural techniques as the organon of cosmotechnics

In this spirit, I propose cosmotechnical inquiry to be always a comparative one, as Hui already seems to adhere to, although for the most part implicitly. Being comparative means to be always aware of the cosmotechnical conditions of the culture to whom the one performing the inquiry belongs, and to relate these to the cosmotechnics proper to the inquired culture. This would lead our attention to the technical activities shaping these cultures' worldviews and to a comparison between their cognitive apparatuses. To a certain extent, I believe this awareness may enable us to avoid the risk of an implicit form of ethnocentrism, that inadvertently projects its own epistemic schemas on its object, claiming at the same time to do the exact opposite, i.e., freeing its object from the cognitive presuppositions of the cosmotechnical *episteme* not belonging to it. A cosmotechnical pluralism is an inquiry in the understanding of technics in different cultures and an analysis of the ways these understandings shape the experience as well as the becoming of the technical systems of these cultures. This attempt becomes really effective only if we admit that to free these understandings from the Western universalising and therefore all-assimilating rationality, allowing them to autonomously develop and reach cosmotechnical self-consciousness, also means to consider the constitutive partiality of perspective of our method of inquiry and the peculiar shape it will give to its results. What characterises this perspective are, once again, the cognitive technologies proper to the culture from which the inquiry is carried out.

Yet which are those technologies and what defines them? As a conclusion for this essay, I will attempt to outline the basic insights of these kinds of techniques, recalling the debate around cultural techniques (Krämer and Bredekamp 20–29) and especially what one of its most representative exponents, Thomas Macho (30–47), claims on this point. If we agree on the fact that every culture is constitutively shaped by its technical media, it should also be fair to admit that not every medium performs a symbolic structuring of the experience. This is precisely what cultural techniques do, in so far as they entail the possibility of their self-representation, which is to say that they enable the one performing them to represent through them the very operation carried out by that technique. As Macho points out,

> [t]he term does not encompass all the techniques a culture has at its disposal, but strictly those techniques that make symbolic work possible. Every culture is grounded in numerous techniques that guarantee its survival [...] Human cultures, however, are not simply composites of these multiple techniques, but evolve out of their symbolic concentration. This symbolic work endows all other activities with their specific meaning; it gives order to the world and enables cultures to develop self-reflexive concepts. Symbolic work requires specific cultural techniques [...] Cultural techniques differ from all other techniques through their potential self-referentiality, a pragmatics of recursion [...] As second-order techniques, cultural techniques have from their very beginning been operating as techniques of self-reflection, identity formation and identification. (30–31)

Thus, there are scriptural, figurative and computing techniques, participating in every culture, significantly differing from one another, but sharing the same recursive potential. The self-representation they enable is the key to access a symbolic dimension, where sense is detached from the empirical schema of experience already subsisting in every technical practice (Di Martino 87–118). By doing so, they allow the members of the culture performing them to practice abstraction and classification and therefore to conceptualise their own technical specificity and possibly the one of other cultures. It is important to notice that although these techniques open up the symbolic and hence linguistic dimension of the human way of life, they are not themselves necessarily symbolic. Their self-representation is just an implied possibility that may, or may not, occur intentionally.[3]

To summarise, as clearly stated by Krämer (157–64), four features can be outlined in order to qualify the specificity of cultural techniques. They perform the unification of the symbolic dimension with the technical, bodily and material components of a culture; as symbolic machines, they enable the exteriorisation of cognitive processes on technical media; in doing so, they allow the one performing them to abstract from the empirical referents of these processes, opening up a symbolic, yet technically embodied, domain; they build up an *episteme*, since their performance always falls in the background, constituting the implicit common frame which renders possible single intentional cognitive operations. As we can see, the reflection on cultural techniques appears to be very close to the field of cosmotechnics, in so far as it tackles the possibility of an embodied, techno-symbolic dimension underlying the manifest cognitive processes endorsed by a culture.

Thus, a cosmotechnical pluralism, aiming to inquire about the differences between the conceptions of technics that cultures develop, without falling prey to Western universalising ethnocentrism, but rather letting these conceptions be conceived starting from the ecological, social and technological background of these cultures, should take into account in the first place the analysis of the cultural techniques adopted by each culture, so that "[t]his is precisely the reason that we have to conceive a cosmotechnical thinking from the standpoint of these ontologies without falling prey to an ethnocentrism" (Hui, "On Cosmotechnics" 8). In the same spirit, the ontological turn in anthropology should be supplemented with a comparative cosmotechnical approach, with special reference to cultural techniques.

This is of the highest importance, because it is only through these special kinds of techniques that it is possible to conceptualise the specificity of our own technical environment and therefore to understand its structural recursion. Indeed, as we have shown above following Hui's insights, humans are characterised by their technical constitution. Yet, the techniques through which they perform this technicity may substantially differ according to each culture. What gives account of this differentiation is the culture's cosmotechnics, i.e., its way of framing experience through technical activities. The latter, in turn, recursively enable the culture to develop one specific cosmotechnics and not another. However, cultural techniques and not technics in general are primarily responsible for the definition of the range of possibilities within which a cosmotechnics can be developed. Furthermore, these particular kinds of techniques are also the ones enabling a culture to explicitly develop its cosmotechnics, that is to say: to conceptualise its own understanding of technics in relation to its environmental as well as institutional conditions, paving the way to the comparative conceptualisation of other kinds of cosmotechnics. A cosmotechnical pluralism, then, should not leave without consideration of a comparison between the cultural techniques that each culture develops and performs.

## disclosure statement

No potential conflict of interest was reported by the author.

## notes

1 It may be of some interest to notice that, different from Hui's approach, Esposito identifies the body and not technics as what enables us to undermine this dichotomy and reconcile what has been illegitimately separated (99–147).

2 It is worth noticing that, incidentally, Descola (*Beyond Nature*) also admits that a collective's technical system can, in turn, sometimes modify its ontology and modes of categorisation (189–96).

3 It may be of some interest here to briefly compare Macho's cultural techniques with what Stiegler (*Technics and Time 3*) calls *hypomnesic* techniques. The latter are those kinds of techniques that actually frame explicitly determinable knowledge, always working as *pharmaka*, that is, in both curative and poisoning way (Stiegler, *What Makes Life*). Although the performance provided by cultural and *hypomnesic* techniques may seem very similar, what distinguishes the former from other, generic technical activities is their self-representative potential, while what differentiates the latter is rather their being susceptible to articulate linguistic knowledge according to a shared criterion of *grammatisation*, that is, both inscription in a medium and agreed understanding of this inscription.

## bibliography

Agamben, Giorgio. "What is an Apparatus?" *What is an Apparatus?, and Other Essays*. Trans. Stefan Pedatella. Stanford: Stanford UP, 2009. 1–24. Print.

Bachelard, Gaston. *Le matérialisme rationnel*. Paris: Presses Universitaires de France, 1953. Print.

Descola, Philippe. *Beyond Nature and Culture*. Trans. Janet Lloyd. Chicago: U of Chicago P, 2013. Print.

Descola, Philippe. "Constructing Natures: Symbolic Ecology and Social Practice." *Nature and Society: Anthropological Perspectives*. Ed. Philippe Descola and Gísli Pálsson. London: Routledge, 1996. 82–102. Print.

Descola, Philippe. "Societies of Nature and the Nature of Society." *Conceptualizing Society*. Ed. Adam Kuper. London: Routledge, 1992. 107–26. Print.

Di Martino, Carmine. "Il simbolismo e i suoi antecedenti." *Sistemi intelligenti* 31.1 (2019): 87–118. Print.

Dobkin de Rios, Marlene, and Ronni Stachalek. "The *Duboisia* Genus, Australian Aborigines and

Suggestibility." *Journal of Psychoactive Drugs* 31.2 (1999): 155–61. Print.

Esposito, Roberto. *Persons and Things: From the Body's Point of View*. Trans. Zakiya Hanafi. Cambridge: Polity, 2015. Print.

Foucault, Michel. *The Order of Things: An Archaeology of the Human Sciences*. Trans. Alan Sheridan. London: Routledge, 2002. Print.

Heidegger, Martin. "The Question Concerning Technology." *The Question Concerning Technology, and Other Essays*. Trans. William Lovitt. New York: Harper, 1977. 3–35. Print.

Holbraad, Martin, and Morten A. Pedersen. *The Ontological Turn: An Anthropological Exposition*. Cambridge: Cambridge UP, 2017. Print.

Hui, Yuk. "Cosmotechnics as Cosmopolitics." *E-flux Journal* 86 (2017): worker01.e-flux.com/pdf/article_161887.pdf. Web.

Hui, Yuk. "On Cosmotechnics: For a Renewed Relation between Technology and Nature in the Anthropocene." *Techné* 21.1–2 (2017): 1–23. Print.

Hui, Yuk. *The Question Concerning Technology in China. An Essay in Cosmotechnics*. Cambridge: Urbanomic, 2016. Print.

Ingold, Tim. *Making: Anthropology, Archaeology, Art and Architecture*. Abingdon: Routledge, 2013. Print.

Kahn, Marvin W., et al. "Australian Aborigines and Alcohol: A Review." *Drug and Alcohol Review* 10.4 (1991): 351–66. Print.

Krämer, Sybille. "Technik als Kulturtechnik. Kleines Plädoyer für eine kulturanthropologische Erweiterung des Technikkonzeptes." *Technik – System – Verantwortung*. Ed. Klaus Kornwachs. Münster: Litt, 2003. 157–64. Print.

Krämer, Sybille, and Horst Bredekamp. "Culture, Technology, Cultural Techniques – Moving beyond Text." Trans. Michael Wutz. *Theory, Culture & Society* 30.6 (2013): 20–29. Print.

Latour, Bruno. *We Have Never Been Modern*. Trans. Catherine Porter. Cambridge, MA: Harvard UP, 1993. Print.

Latour, Bruno, and Steve Woolgar. *Laboratory Life: The Social Construction of Scientific Facts*. Beverly Hills: Sage, 1979. Print.

Leroi-Gourhan, André. *Gesture and Speech*. Trans. Anna Bostock Berger. Cambridge, MA: MIT P, 1993. Print.

Macho, Thomas. "Second-Order Animals: Cultural Techniques of Identity and Identification." Trans. Michael Wutz. *Theory, Culture & Society* 30.6 (2013): 30–47. Print.

Moore, Gerald. "Dopamining and Disadjustment: Addiction and Digital Capitalism." *Are We All Addicts Now? Digital Dependence*. Ed. Vanessa Bartlett and Henrietta Bowden-Jones. Liverpool: Liverpool UP, 2017. 68–75. Print.

Simondon, Gilbert. *On the Mode of Existence of Technical Objects*. Trans. Cecile Malaspina and John Rogove. Minneapolis: Univocal, 2016. Print.

Sloterdijk, Peter. "The Domestication of Being: The Clarification of the Clearing." *Not Saved. Essays after Heidegger*. Trans. Ian A. Moore and Christopher Turner. Cambridge: Polity, 2016. 89–148. Print.

Sloterdijk, Peter. "The Time of the Crime of the Monstrous. On the Philosophical Justification of the Artificial." *Not Saved. Essays after Heidegger*. Trans. Ian A. Moore and Christopher Turner. Cambridge: Polity, 2016. 281–95. Print.

Sloterdijk, Peter. "Wozu Drogen? Zur Dialektik von Weltflucht und Weltsucht." *Weltfremdheit*. Frankfurt am Main: Suhrkamp, 1993. 118–60. Print.

Stiegler, Bernard. *Technics and Time 1. The Fault of Epimetheus*. Trans. Richard Beardsworth and George Collins. Stanford: Stanford UP, 1998. Print.

Stiegler, Bernard. *Technics and Time 3. Cinematic Time and the Question of Malaise*. Trans. Stephen Barker. Stanford: Stanford UP, 2010. Print.

Stiegler, Bernard. *What Makes Life Worth Living: On Pharmacology*. Trans. Daniel Ross. Cambridge: Polity, 2013. Print.

Viveiros de Castro, Eduardo. *Cannibal Metaphysics*. Trans. Peter Skafish. Minneapolis: Univocal, 2014. Print.

andrés vaccari

# NEOSUBSTANTIVISM AS COSMOTECHNICS

## gilbert simondon versus the transhumanist synthesis

### I neosubstantivism, cosmotechnics, technocene

Technological substantivism, a philosophical view of technology long considered defunct, has acquired a renewed and urgent relevance. I will refer to *neosubstantivism* as the contemporary, mutant version of substantivism that underlies dominant visions of technology. In turn, neosubstantivism designates our present *cosmotechnics* (Hui, *Question*, "Cosmotechnics as Cosmopolitics," and "On Cosmotechnics"), whose other name is the *Technocene* (Sloterdijk 328).

Neosubstantivism has colonized our politics (left, right, and center) to become the ruling theology of our end times, complete with a myth of Creation in which humans are born from a "technogenesis" (Stiegler 26–27; Hayles 1–18) and an eschatology in which the species transcends the flesh to become One with the Godhead of technology.[1] The church of neosubstantivism has left-wing accelerationists, alt-right apocalyptic antihumanists, zealot Singularitarian entrepreneurs, technophile cornucopians, and even ecologically inclined degrowthers, praying at the same altar. The blockbuster success of this religion in its various political orientations is partly due to the fact that it articulates a palpable and prevalent feeling in industrialized societies: the notion that there is no *outside* of the global system. Mark Fisher has famously referred to this feeling as *capitalist realism*: "the widespread sense that not only is capitalism the only viable political and economic system, but also that it is now impossible even to imagine a coherent alternative to it" (2). As the extractive capitalist machine reaches its material limits, we feel we are trapped inside a vehicle with malfunctioning brakes, racing towards the edge of a gaping precipice. The experience of the "system" as a monolithic Behemoth leads to an awareness of planetary boundaries (Nitzke and Pethes) and of the Earth as a *hyperobject* (Morton). Once their mechanisms are set in motion, systems obey the law of inertia and move with a force proportional to their mass. There is no alternative because capitalism is *substantialized* in a vast web of machines and structures with their immovable trajectories in which humans are embedded. This resignation, the inability to dream outside the system, soaks and corrodes our political imaginary.

Beneath their differences, all variants of neo-substantivism have this central belief in common: the overarching and all-pervasive civilizational role of technology in the future of the human species, a cosmic role previously afforded to divine or natural forces. I will focus on the myth of the *Singularity* as a case study on neosubstantivism, found here in the context of an apologia for free market economics and conservative politics.[2] Singularitarianism is worthy of study because of its ideological power, its representativeness, and because of the peculiar way in which it naturalizes technology. I hope to mobilize the philosophy of Gilbert Simondon and the conceptual resources of Yuk Hui's cosmotechnics to disassemble some central assumptions of Singularitarianism. These two philosophical alternatives offer lines of flight out of the Technocene by questioning the neosubstantivist identification of technology and cosmology.

The naturalization of technology[3] as thing and process, and its expansion to the status of all-encompassing cosmic force are two aspects of the same shift. Questioning the anthropocentric bias that underlies the notion of an "Anthropocene," the Technocene emplaces technology as the true subject of history, a force which is not only akin to nature but which comes to swallow and replace nature as a "neo-environment" and *autotelic totality* (Cera 244). As Hui remarks, the order is reversed: from playing a supplementary and prosthetic role, technology becomes the origin, "itself the ground in contradistinction to the figure" ("On Cosmotechnics" 320). Another way to see the Technocene is as techno-fetishism, which Benjamin Noys describes as a "mysticism of [the] material object being treated as possessed of divine powers [...] The result is the inflation of the technological object to something that horrifies and fascinates, electing it out of history into a natural or metaphysical realm" (3). The deomorphic role of technology is patently visible in everyday discourse, where technology is routinely portrayed as a substantial subject with omnipotent capacities. A quick Internet search yields familiar statements such as: *Technology will change the way we live*, *5G wireless technology will muck up weather predictions*, *Technology will not save us*, *Digital technology will strengthen America's biggest banks*, *How technology will change the clothes we wear*, etc. The agential force of technology as subject has the ideological function of occluding the real forces at work behind technological change.

We thus find ourselves in the thicket of two founding theses of classical substantivism. The first thesis is the one substantivism is most renowned for: technology is an autonomous entity that expands following its own internal, self-given laws in a manner wholly opaque to human agency and control. Jacques Ellul, the French Catholic thinker most readily associated with substantivism, disaggregates the systemic, quasi-organismic nature of the "technological system" into five defining features: *automatism*, *monism*, *self-augmentation*, *universalism*, and *autonomy* (*Technological Society* 79–148).[4] The global technological system is a collection of tightly interrelated technologies ("techniques," in Ellul's parlance) that "combine to form a whole, each part supporting and reinforcing the others" (111). Hence, in Ellul's terms, technology "has taken substance, has become a reality in itself. It is no longer merely a means and an intermediary. It is an object in itself, an independent reality with which we must reckon" (63).

It follows that it is futile "to hope to be able to suppress the 'bad' side of technique and preserve the 'good'" (Ellul, *Technological Society* 111). *Value-ladenness* is another founding thesis of substantivism: technology exceeds the merely "technical" to embody values that inform its very material configuration. Artefacts, machines, and techniques are not ontologically neutral and morally innocent vehicles for the expression of human goals; on the contrary, values are substantialized in material structures and the interactions they enable. In riding a car or drinking from a Styrofoam cup, we are assenting to values we may have not chosen and which many of us are probably unaware of. From a wider perspective, a given technological configuration can incarnate a whole epoch or *Geist*.

According to this view, it may be legitimate to conflate capitalism, its technologies and the Technocene as facets of the same irreducible epochal phenomenon.

However, the autonomous force of technology also escapes the intentions of the ruling socioeconomic and political order. The "automatism of technique" will end up crushing capitalism, which is unable to control the forces of production and prevent them from turning against it (Ellul, *Technological Society* 82). Ellul cites approvingly from *The Communist Manifesto*:

> Modern bourgeois society with its relations of production, of exchange and of property, a society that has conjured up such gigantic means of production and of exchange, is like the sorcerer who is no longer able to control the powers of the nether world, whom he has called up by his spells. (Marx and Engels 41)

On their part, accelerationists believe it is possible to repurpose the machines of capitalism and put them to other ends, in the context of an alternative post-capitalist society (Srnicek and Williams). One important question is whether these structures can be co-opted or redirected towards other ends.

It follows that the Technocene displays a polarity: the *near* and the *far*. While the *far* reaches out to the blurry spatio-temporal boundaries of a whole epoch, the *near* points to the technological shaping of actions and perceptions; the *pragmatic* and *hermeneutic* dimensions of mediation, respectively – a topic of central interest to post-phenomenological accounts of technology and design, such as Peter-Paul Verbeek's (*What Things Do*, "Materializing Morality," and "Obstetric Ultrasound") and to Actor-Network Theory (Latour, *Science in Action*, *Pandora's Hope*, and *Reassembling the Social*). These views converge on the nodal insight that artefacts can display quasi-intentionality and goal-directedness, materializing "scripts" (Latour, "On Technical Mediation" 31) and "programs of action" (Akrich and Latour 260–61) that exert a constitutive, causal force on the shaping of human intentions and actions – what we may call *microsubstantivism*.

In this context, Hui's concept of *cosmotechnics* can fulfill at least three important critical functions. Firstly, the notion of cosmotechnics returns technology to history – or better still, to *histories*. To return technology to history may help us break the stronghold of neosubstantivism and challenge modernity's hold on technology. Secondly, cosmotechnics may help us think beyond the traditional distinction between nature and technology without collapsing the former into the latter. Thirdly, cosmotechnics highlights moral aspects of technically mediated practice and restores some space for human agency; this is important inasmuch as any cosmotechnics is also a *cosmopolitics*.

Cosmotechnics stands for the deep interweaving of human action and technology as shaped by diverse moral universes (Hui, *Question* 18–32). Cosmologies provide "not only schemas that define the modes of participation, but also correspond to the moral grounds of such participation" (Hui, "On Cosmotechnics" 321). Cosmotechnics is plural, an invitation to dream alternative futures, another technology and politics beyond the Technocene. Hui's work focuses mainly on Chinese cosmotechnics with a view to reopen "the question of technology through the affirmation of non-modern cultures" (338). Our strategy is to stay within Western philosophy and approach cosmotechnics from the perspective of Simondon. As we shall see, the Singularity preaches passivity before a cosmos ruled by deomorphic technology. Simondon, on the other hand, regards the relation between humans and their creation as one of inter-individual collaboration. Next, we shall delve deeper into these contrasting views.

## 2 the singularity and the transhumanist mutant synthesis

The Singularity is a particularly virulent form of neosubstantivism that provides an image of technological change friendly to the tech-industry and the conservative political agenda. What makes Singularitarianism worth examining is its performative aspect, the fact it is fast becoming a self-fulfilled prophecy. It is a belief that drives technological change itself.

The Singularity is an End of History narrative that predicts a future point in time where technological progress accelerates to infinity (Ellul's self-augmentation). Although originally the Singularity only concerned machine intelligence (Vinge), the engineer and futurologist Ray Kurzweil broadened its scope to include all information technologies and related developments. "The key idea underlying the impending Singularity is that the pace of change of our human-created technology is accelerating and its powers are expanding at an exponential pace" (Kurzweil, *Singularity* 7–8). Kurzweil has turned the Singularity into a successful business model, establishing it as the core philosophy for a tech-startup "accelerator" located at Singularity University, a venture he started in 2008 with fellow businessman and Singularitarian Peter Diamandis. In Kurzweil's view, the Singularity does not merely entail that machine intelligence will surpass human intelligence, but that there will be a merging of the two that will mark the entry into a post-biological mode of being:

> The Singularity will allow us to transcend these limitations of our biological bodies and brains. We will gain power over our fates. Our mortality will be in our own hands. We will be able to live as long as we want [...] By the end of this century, the nonbiological portion of our intelligence will be trillions of trillions of times more powerful than unaided human intelligence. (*Singularity* 8)

Our descendants will lead a digital existence on a substratum of nanobot clouds. Their exodus from Earth in search of intergalactic adventure "is the ultimate destiny of the Singularity and of the universe" (Kurzweil, *Singularity* 21). The progression leading to this point is presented as an innocent induction from historical trends, hinging mainly on the rapid increase in computational power. For Kurzweil, technological change is driven by natural forces that are an unproblematic extension of "evolutionary" trends: "Exponential growth is a feature of any evolutionary process, of which technology is a primary example" (11). Evolution seeks to create patterns of ever "increasing order" (13) in a process that will eventually lead to the Sixth Epoch of Evolution, where intelligence "will begin to saturate matter and energy in its midst" (21).

The central strategy of the Technocene is the *naturalization* of technology, sometimes in a theological or cosmic guise. In Kurzweil's version, naturalization is the ground for a *moral legitimation* of free market economics that continues a tradition inaugurated by the Physiocrats in the eighteenth century, and later immortalized in Adam Smith's famous metaphor of the Invisible Hand. The naturalization of the markets was partly a product of the spread of the mechanistic worldview into social phenomena. As an extension of mechanical nature, society was conceived in terms of a social physics and the market as a self-regulating machine that should remain free of all political intervention. In a book called *Bionomics: Economy as Business Ecosystem*, Michael Rothschild states the thesis clearly:

> Capitalism, or the market economy, or the free-enterprise system – whatever you choose to label it – was not planned. Like life on earth, it did not need to be. Capitalism just happened and it will keep happening. Quite spontaneously. Capitalism flourishes whenever it is not suppressed, because it is a naturally occurring phenomenon. It is the way human society organizes itself for survival in a world of limited resources. A capitalist economy can best be comprehended as a living ecosystem. (xi)

"Networks" are the present descendants of self-regulating machines such as the balance, which provided inspiration for the first wave of mechanistic models of the market (Mayr). The Singularity advocates the deregulation of markets and a "proactionary" approach to technological development that is also championed by other transhumanists such as Max More and Fuller and Lipinska.

On his part, Kevin Kelly, founder editor of *Wired* magazine and another Silicon Valley visioneer, follows this same path, coining his own neologism, the "technium," to refer to the global network of machines and systems that constitutes a global organism in its own right (15). Kelly presents this uncredited repackaging of Ellul's *technique* in a book entitled *What Technology Wants*. This title boldly illustrates the technology-as-subject at the heart of the Technocene. Opting for vitalist and animistic metaphors, Kelly refers to the technium as the Seventh Kingdom of Nature, an organic whole that evolves following the same principles as living things:

> [...] systems – all systems – generate their own momentum. Because the technium is an outgrowth of the human mind, it is also an outgrowth of life, and by extension it is also an outgrowth of the physical and chemical self-organization that first led to life. The technium shares a deep common root not only with the human mind, but with ancient life and other self-organized systems as well. (15)

In classical substantivism, technology and nature remain ontologically distinct, and the former is often seen as exploiting, subjugating, or mediating the latter. "Technique is opposed to nature," writes Ellul. "Art, artifice, artificial: technique as art is the creation of an artificial system" (*Technological Society* 79).[5] For neo-substantivism, this distinction has ceased to hold any meaning. The central irony here is that substantivism started out its philosophical career as a cautionary view on the catastrophic effects of modern technology, and now Silicon Valley's think-gurus have co-opted and retro-engineered it into a technophilic fable. If there is any "original" nature left somewhere, it doesn't matter anymore. As Zoltan Istvan, US presidential candidate for the Transhumanist Party, claims:

> What we're doing to the planet is not as important as what we are achieving as a species entering the transition to the transhumanist age [...] [E]nvironmentalists are mistaken in thinking the Earth is our only or permanent home. Before the century is out, our home for much intelligent life will likely be the microprocessor. We will merge with machines and explore both the virtual and physical universe as sentient robots. That's the obvious destiny of our species and the coming AI age.

The fantasy of flight from planet Earth grotesquely reimagines capitalist realism in terms of a parasitic global organism that, after sucking its cradle planet dry, sets out on a mission to lay waste to other worlds. In this manner, the Singularity narrative pushes out the limits to capitalistic growth to the far reaches of the cosmos. The main message is that we should not worry about altering current power structures or seek less self-destructive socioeconomic arrangements. According to Kurzweil, the exponential logic of technological growth is negentropic and follows a Law of Accelerating Returns ("Law of Accelerating Returns" 383). From this angle, the Singularity is a version of technophile cornucopianism, also known as the *image of the unlimited good*, or the *open systems view* (Trawick and Hornborg 3).

The moral function of a cosmotechnics is to regulate human behavior in the context of an overarching and self-contained image of the world. A cosmotechnics legitimates certain actions and opinions firstly by shaping the perceptions of those that subscribe to it. In this case, the function of the Singularity is to provide an optimistic picture of technology and sanction the central role of the entrepreneur as the motor of history. It provides a holistic image to counter the disorientation often felt in the face of complex, traumatic change.

To begin with, Singularitarianism has been designed to be broad and vague enough to accommodate any new technology released into the market and enthusiastically endorsed in the media. Nanomedicine, wearable tech, cognitive implants, expert systems, *in vitro* meat – anything can be incorporated into the story. The function of Singularitarianism is to cushion the low points of the tech hype-cycle, promoting optimism and continued investment in the face of market downturns and cultural backlashes. Ensuring this hype-sustaining effect requires an adequate temporal frame. As any self-respecting prophet knows, your prediction must be neither set too *early* (in case it is promptly defeated by facts) nor too *late* (ensuring your audience continues to believe they may enjoy the promised benefits in their own lifetime).

In the threshold of the Sixth Epoch of Evolution, Silicon Valley cosmotechnics promotes an attitude of hopeful passivity before the monumental forces of the Seventh Kingdom of Nature – unless you are one of the few powerful actors granted a significant role in this play. Singularitarianism represents technological acceleration as a spontaneous process, free of control and from vested human interests, but the reality is the opposite: the Singularity promotes a narrow view of innovation driven by technologies of control, and managed by a few powerful (mostly) men – to the extent that technology *can* be "managed" in the Technocene. It is a worldview tailor-made for and by the military–digital complex. Here, the locus of power is the commercial and proprietary control that the capitalist system demands from all "innovation."

Adapting substantivism to the worldview of the tech entrepreneur requires some fine-tuning, beginning with the introduction of substantivism's traditional foes, *instrumentalism* and *determinism*, into the mix. In their seminal essay "The Californian Ideology," Barbrook and Cameron already draw attention to the "contradictory mix of technological determinism and libertarian individualism" that characterizes the orthodoxy of Silicon Valley's virtual class (54). Determinism provides the link between substantivism, instrumentalism, and individualism. Briefly stated, technology follows its own developmental path (substantivism) *and* is the cause for change in all other spheres of human activity (determinism); in turn, this means that the individual in charge of introducing new technologies into the market occupies a privileged place at the center of the cosmos, as the unmoved mover behind all historical change. At some point in this story, instrumentalism is introduced: the view that technologies are neutral vehicles for human intentions, making no causal contribution to the outcomes of action, let alone to the genesis of those very intentions. Advertisements and corporate mission statements issuing from Silicon Valley make routine use of instrumentalist narratives in which the user is positioned as a free, empowered individual who uses technologies to accomplish his or her goals more effectively and broaden the horizon of agential possibilities. Finally, we should add another *-ism* to the picture: *solutionism*, the view that technology can solve all of humanity's problems (Morozov).

Singularitarianism pushes instrumentalism to its limit, and in a very peculiar way. The Singularity marks the point of convergence between technological evolution and the spiritual ascension of a chosen group of humans. Technology becomes the vehicle for spiritual transcendence, the instrument for the fulfilment of human nature. The logical end point of the autonomous evolution of technology coincides with the consummation of the cosmic destiny of humanity. Both moments are one and the same. In this way, technology not only mediates but substantializes the union of humanity and the universe in a sort of Holy Trinity (human–machine–cosmos). To this end, Kurzweil claims that humanity's nanobotic descendants will preserve their human essence. In this sense, Kurzweil's version of the Singularity is humanistic and somewhat reassuring.

Singularitarianism draws transhumanism into its fold by way of this numinous narrative, to give birth to what I call the *Transhumanist Synthesis*: the terminal merger of Singularitarianism with the trope of "deliberate selection" and "enhancing evolution" through market-based eugenics (Harris 11). The Transhumanist Synthesis emerges partly from the failure of previous forms of "soft," humanistic transhumanism; and partly from the voracious colonizing effect of the market itself. Soft transhumanism is grounded on the idea of wide beneficence. "The claim is that for *most* current human beings, there are possible posthuman modes of being such that it could be good for these humans to become posthuman in one of those ways" (Bostrom, "Why I Want" 108; emphasis in the original). In the original view of Directed

Evolution, posthumanity will be the cumulative result of individual enhancements on species-wide traits. Thus, a better humanity will arise from the action of the Invisible Hand. The human enhancement project appealed to a future state of collective beneficence that would affect all or most of humanity for the better. The writings of Nick Bostrom (e.g., "Human Genetic Enhancements," "In Defense of Posthuman Dignity," and "Why I Want to be a Posthuman"; Bostrom and Sandberg, "The Wisdom of Nature") represent the most developed attempts at a philosophical defense of this earlier stage of transhumanism and clearly lays bare its failure (see Vaccari, "Why Should").[6]

The Synthesis abandons all pretense of beneficence to relentlessly insist on the inevitability of technological "advance," on the *necessity* of what technology wants. The appointment of Kurzweil as Head of Engineering at Google sealed the creation of the Synthesis by absorbing transhumanism into Silicon Valley's ideological agenda. Posthumanity will not necessarily be better, not even good. Naturally, this means that the "Rapture for nerds" (MacLeod 115) will be for an elected few, while the rest of our descendants will have to be content with remaining Mostly Original Substrate Humans (Kurzweil, *The Age* 229) – assuming they don't perish in the post-capitalist apocalypse left behind by the cloud of nanobots.

## 3 simondon against silicon valley

Our final step is to examine the thought of Simondon from the perspective of cosmotechnics, mining philosophical resources towards a critique of neosubstantivism in the guise previously examined. What does Simondon have to say to the ruling techno-theocracy?

In the introduction to *On the Mode of Existence of Technical Objects* ((1958) 2017), Simondon sets a task for the philosopher, that of the critic of myths. Simondon is concerned with bridging the distance between culture and technology that arises partly as a result of fragmentation and hyper-specialization (118). Culture treats the technical object as foreign, inhuman, and contradictory: as a lump of matter *and* as something that harbors intentions, good or (mostly) evil. Culture does not understand technology because it considers the technical object as identical to the *machine*: a closed block, an impenetrable black box. The myth of the *robot* encapsulates the alienation of the technical object in modern culture, combining fantasies of power while substantializing and materializing the technical object (157).

To remedy this state of affairs, Simondon seeks to establish a pedagogical program and political project, a *new encyclopedism*, and hence a *new humanism*, that can fully integrate the technical object into culture. For Simondon, humanism is primarily concerned with human liberation, and the goal of freedom is achieved with a universal kind of knowledge. Simondon's account of technology aims to free human beings by opening up the technical object and elucidate its universal mode of existence.

> [The human] is enslaved to his dependence on unknown and distant powers that direct him while he can neither know nor react against them; it is isolation that enslaves him [...] Having become a machine in a mechanized world, he can regain his freedom only by taking on this role and by superseding it through an understanding of technical functions thought from the point of view of their universality. (Simondon, *Mode of Existence* 117)

Previous to the nineteenth century, technology was grounded on bodily and perceptual schemas. From a phenomenological and social perspective, humans perceived themselves to be at the center of technical action. However, in the industrial revolution, the individual becomes a "mere spectator" of the machines to work (Simondon, *Mode of Existence* 132). Thus, the alienation denounced by Marx is as much economic as it is physiological and psychological (133). From this alienated point of view, progress is no longer lived as a continuous process, but as a fragmented one, proceeding in leaps and bounds. The resulting image looks suspiciously current:

> Progress is henceforth thought of as cosmic, at the level of its overall results. It is thought abstractly, intellectually, in a doctrinal manner. Progress is no longer thought by craftsmen, but by mathematicians, who conceive of progress as man taking possession of nature. Beginning with the Saint-Simonians, this idea of progress starts to support technocratism. (Simondon, *Mode of Existence* 132)

Simondon's theory of technology follows seamlessly from his theory of *individuation*. Individuation is the thread that connects Simondon's analysis of physical systems, living things and technologies.[7] The distinction between these three realms is not due to an a priori ontological classification that picks out essences, identities, or substances, but which focuses on operations. In Simondon's genetic and processual view, nature, the human, and technology are marked by different *regimes of individuation*. In this scheme, technical objects follow different laws than those of biological evolution, although they *tend towards* natural objects and may even lose their artificial character in some cases (Simondon, *Mode of Existence* 49). *Concretization* denominates the specific regime of individuation proper to technology, a process of functional and structural convergence in which a technical object progressively gains complexity by "informing" itself. The technical object develops following internal norms (structural, energetic, and functional) and in phases oriented towards greater perfection. In this process of "information" (a term which Simondon defines in his own specific way), preindividual potentials are actualized into structures.

Although, on a first approach, Simondon's mechanology appears strongly committed to substantivism, we should follow Darío Sandrone in labelling this philosopher a *technical realist*, in the context of a tradition of machine theory that encompasses Karl Marx, Robert Willis, Franz Reuleaux, and Jacques Lafitte. Divided into weak and strong currents, the tradition of technical realism understands technology as a form of knowledge "constituted by actions and objects that, to a large extent, are linked to the objective principles of the natural sciences and mathematics, and which possess certain autonomy with respect to the practical function determined by the context of use" (Sandrone 309–10; my translation).

The dynamics of technological change can enter into conflict with economic pressures, as well as with cultural and social standards. Whereas for Singularitarianism the dynamics of capitalism are perfectly integrated with the "natural" course of technological progress, for Simondon the capitalistic values that drive production and consumption often hinder the immanent logic of the technical object. The market constrains and slows down technical becoming by adding contingent, superfluous structures and forcing extraneous functions on the "pure" technical object. Conversely, the way to rectify our alienation and achieve true "progress" is not through the mechanisms of the free market, less through some search for technological transcendence, but through a concerted and collective design of systems beyond the narrow concerns of certain socioeconomic sectors. Simondon's cultural revolution is meant to be broad and far-reaching, a true reform of culture with technology at the center.[8] The destiny of the human is, for Simondon, intimately tied in with technology, but Simondon sees this relation as one of collaboration and inter-individual coupling. Likewise, Simondon's philosophy resists the mechanistic thesis at the heart of the transhumanist project. Human beings are not machines. Technology originates in life, in the struggle for dominance over the environment. Technical invention issues from the kind of problems that living things face in their everyday struggle for survival. Invention arises from interruptions and discontinuities that challenge the organism's "operative accomplishment continuous with its project [*accomplissement opératoire continu dans son projet*]" (Simondon, *Imagination* 139). The living dimension is the key to understanding the relation between humans and technologies, and the function of humans in technical ensembles.

> There is something alive in a technical ensemble, and the integrative function of

> life can be ensured only by human beings; the human being has the capacity to understand the functioning of the machine, on the one hand, and the capacity to live, on the other: one can speak of technical life as being that which actualizes this relation between these two functions in man. Man is capable of taken upon himself the relation between the living being that he is and the machine he fabricates; the technical operation requires both technical and natural life. (Simondon, *Imagination* 140)

The passage in *Mode of Existence* describing the complementarity of human and machine memory is illustrative of how the collaboration between humans and machines works (Simondon, *Mode of Existence* 135–39). There is no isomorphism or equivalence between the functions carried out by technical and living things. It follows that, for Simondon, it would be impossible to "download" consciousness onto an artificial substratum, since mental life is the outcome of very specific structures and processes; in this sense, Simondon is a *body-centrist* (Clark 56). The concreteness of biological beings, the complex web of reciprocal causality upon which their performance depends, sets a material limit to functional substitution and to potential mergers.[9] Likewise, paradigms such as automation, standardization, augmentation, and enhancement fail to grasp the true relation between humans and technical objects. In Simondon's terms, the function of the Singularity is to produce a "technophany" (*technophanie*), a ritualistic and symbolic guise that veils the technical object; it allows the insertion of the object into culture, yet it immobilizes it, impeding its free expression ("Psychosociologie de la technicité") (*Sur la technique* 39–52).

The cosmotechnical nature of Simondon's thought is most visible in its *normativity*. The starting point for a Simondonian cosmotechnics is the recognition of the specific dynamics of technical reality and its role in transindividuation. Simondon demands that we establish an *ethical* relationship with the technical creatures in our midst. Modern cosmotechnics is alienating precisely because it degrades technical objects, stripping technical action from any ritual and moral overtones, and framing it as merely mechanical, instrumental action. Rather than a tool or a utensil, the technical object is

> [...] condensed human effort, waiting, an available virtual being, a potential action. For this reason, we must not only refine our gaze, to purify it, but we must also reform the technical operation: it must aim to constitute an open object, perfectible, and neotenic, that is to say depository of an evolutionary potential; this object must not be a sold, possessed thing, but a thing that institutes a participation [...] [F]ree in relation to the human, though linked to him, the technical object augments the density of the human field of activity: it is really a social being [...] the concrete ensemble [*l'ensemble concret*] is the couple human–machine. Modes of being and of thought, types of social structure that could not exist with individuals without equipment, solitary humans, can be concretized and structured through this contribution, this enrichment in potential. (Simondon, *Sur la technique* 363; my translation)

A cosmotechnics is an *image* in the Simondonian sense: a pattern or schema that propagates transductively through the social body and which underpins its collective activity, helping to structure it (Bardin 222). At the heart of this image, we find the duality of human action, which presents at once a *technical* and a *symbolic* aspect (Bardin 177–78). Simondon is a thinker of cosmotechnics par excellence, since his philosophy unites the image of the cosmos, the symbolic and the technical as three aspects that: (1) are condensed in a single gesture; and (2) point to the common origin of all forms of human thought in a primordial relation to the cosmos.

In Simondon's account, technical thought splits off from a primordial mode of grasping of the world which progressively unfolds into different *phases*, in the strictly physical meaning of the term (*phase ratio*).[10] The source of magical thinking is a basic structure of perception, the *figure-ground*, one step "immediately above the relationship that is simply between the living thing and its milieu" (Simondon, *Mode of Existence* 169). Magical thought grasps the world as a network of meaningful spots in the landscape: a mountain, a cave, events such as thunderstorms. It does not separate objects and individual humans. As Hui claims, in the magical phase, there is "hardly any distinction between cosmology and cosmotechnics, since cosmology only makes sense here when it is part of everyday practice" (*Question* 21). Technical, aesthetic, and religious modes of thought originate "at the moment of the splitting of the primitive magical unity" (Simondon, *Mode of Existence* 174). With technical thinking, the figure is decontextualized from the background and the key points are reconceptualized in functional terms. The powers and forces that were, until then, qualitative and concrete become mobile, and objectified in tools and instruments. Aesthetic thought appears as a product of this split, rather than a phase; it is "a permanent reminder of the rupture of unity of the magical mode of being, as well as a reminder of the search for its future unity" (174). Technical thought moves forward by multiplying objects "without being able to recapture the primitive unity" (186). Fragmentary and multiple by nature, technical objects engage the world at localized and particular points.

Now, if human cultures carry on their business against the ground of an independent technical reality, is it possible to have *multiple* cosmotechnics? Or do all cultures converge towards a single image of technics, determined by a universal *tendency*, to use Leroi-Gourhan's term (90)? Simondon sees no contradiction in the plurality of cosmotechnics and the universal paths of technical objects. On the one hand, there is a *real* technical development beyond cultural dynamics. On the other, there is a rich proliferation of techniques and ways of doing things, inserted in an equally diverse variety of cosmologies and symbolic practices. Simondon opens a middle ground between the universalizing, profane cosmotechnics of Western metaphysics, and the notion of a cultural constructivism unhinged from any material constraints. Technics is a *limit* that remains invariant across cultures, yet it can be overwritten by many different significations and symbolic charges, owing to the common origin of technics and symbolic thought that subsists in human gesture. In fact, Simondon goes further, arguing that the distinction culture-technics is internal to technics itself (Bardin 185).

> It would therefore be more correct not to use the word *technique* to contrast it with *culture*: "culture" and "technique" are both activities of manipulation, therefore techniques: they are even techniques of human handling, because they exert an action on man through the intermediation of the environment in the case of activities generally called technical, and directly in the case of culture. (Simondon, *Sur la technique* 318)

The technical object is not outside of culture but at the center of it, as the collective depository of cognitive schemas and norms of action (Bardin 197–98). The technical object carries the residual force of life at the heart of the act of invention. The human being, and the living being more generally, is essentially a *transducer*, a receiver and transformer of information (Simondon, *Mode of Existence* 155). As such, the human is the "permanent organizer," "the living interpreter of all machines among themselves." Machines need the human "in the same way musicians in an orchestra need the conductor" (17). Technics establishes a communication between humans, diachronic and synchronic, by opening up a shared world of possible actions and by sedimenting what Bernard Stiegler calls "tertiary retentions" (1). In this sense, technics is always already anthropotechnics. As the substratum for transindividual processes, technical objects are political players that produce structural effects in the form of social systems. According to Andrea Bardin (221), for Simondon, the political is an ontogenetic process that runs through the social body, where social systems emerge and develop.

## 4 conclusions

The point of the preceding exercise is to help loosen the grip that neosubstantivism has on our technological imaginary, drawing from the resources of the Western tradition. I hope to have hinted at how Simondon can help us build a liberating, alternative picture of the relationship between humans, technical things, and the cosmos. Likewise, the cosmotechnical view, as proposed by Hui, splinters the monolithic hold of *a* technological system. Systems, especially one as vast and fragile as the technological system, are fundamentally open to contingency and plurality (Hui, *Recursivity*).

Should Silicon Valley read less Harari, and more Simondon and Hui, then? The main obstacle to establish some kind of dialogue or dialectic is communicational. It has to do with the present state of culture, where the transmission of information and ideas is increasingly segmented in terms of bubbles and resonance chambers. Since all "metanarratives" have long been pronounced obsolete (Lyotard 34–52), perhaps a more apt term for a narrative such as the Singularity is "megadiscourse" (Winner, "Decadencia y caída del tecnotriunfalismo" 127), a metanarrative divorced from all notion of truth. The function of a megadiscourse remains ideological (the legitimation and reproduction of existing power structures) but a key distinguishing feature is that it is not meant for universal agreement and consumption. A megadiscourse is addressed to a limited audience of converts and its function is to prop up bubbles: market bubbles, cultural bubbles, ideological bubbles. Megadiscourses are manufactured just like any other product. In the 1940s, Ernst Cassirer observed that the engineering of myth is a phenomenon peculiar to the twentieth century. Myths, according to Cassirer, are no longer the spontaneous products of the collective imagination but "artificial things fabricated by very skilful artisans." Our politicians and manipulators of perception have developed a "new technique of myth" that enable myths to "be manufactured in the same sense and in accordance with the same methods as any other modern weapon" (Cassirer 282), a fitting analogy in the context of the military–digital–industrial complex of which Silicon Valley is a key ally.

Further, Cassirer relates the phenomenon of myth-manufacture to a shift in the function of language, which returns to its magical, performative roots. Modern myths rely on magic words that are "destined to produce certain effects and to stir up certain emotions [...] new-fangled words [...] charged with feelings and violent passions" (Cassirer 283).

The problem, then, may well be one of information flows, of open vs. closed systems. Megadiscourses can only emerge in closed systems, such as Silicon Valley's cultural bubble. As Joichi Ito argues: "In Silicon Valley, the combination of groupthink and the financial success of this cult of technology has created a positive feedback system that has very little capacity for regulating through negative feedback." The task of philosophy is to provide a vantage point that cuts across and transcends all perspectives on the technical object: that of the scientist, the engineer, the manufacturer, etc. Further, philosophy is a reflexive activity that feeds back into the technical and symbolic milieu that supports the functioning of social systems. As a transindividual activity, thought is reflexive "insofar as it can change the cultural milieu from which it emerges" (Bardin 202). However, as Bardin suggests, Simondon seems overly optimistic here, as his project requires strong institutions that can mediate the feedback loop between technics and transindividuation; those are, precisely, the institutions that free market mechanisms aim to weaken with a view to eradicate.

## disclosure statement

No potential conflict of interest was reported by the author.

## notes

1 Theology is the study of the divine and of religious belief. I use the term in two more specific senses. Firstly, as an account of the origin and destiny of human creatures in the context of a meaningful cosmos either created by God or rendered rational by the future arrival of a God to

come: the reign of "Homo Deus" (Harari), or the human rendered divine by technology. In the second sense, theology refers to political theology, following the philosophy of Carl Schmitt: the way theological narratives and structures consolidate into earthly power structures. In this last sense, theology is "the study of the structures and sources of *legitimacy* – of the ways that people attempt to answer the question of who should be in charge and why" (Bailes).

2 We should keep in mind that the Singularity also features in Nick Land's "dark accelerationism" as the anti-humanist version of Kurzweil's cozy, humanist narrative (see Land, "Meltdown" and "Ideology, Intelligence, and Capital").

3 The term *technology*, as opposed to *technics*, has a number of interpretations. I will use technology to also encompass technics in general, except in certain passages where drawing the distinction becomes crucial to the argument. Following Simondon, I will employ technology to refer to: (1) the modern epoch in the history of technics; and (2) a structured and universal body of knowledge on technics.

4 Ellul published his major work, *La technique ou l'enjeu du siècle* (*The Technological Society*) in 1954, an *annus mirabilis* for technological substantivism, since it also saw the appearance of Martin Heidegger's *Die Frage nach der Technik* (appearing in *Vorträge und Aufsätze* (Pfullingen: Neske, 1954)). This work of Heidegger is also considered to advance a strong substantivist view of technology, although based on very different premises than those of Ellul. Due to reasons of space, we will not get a chance to discuss Heidegger here.

5 When Ellul denounces the encroachment of technology on nature, and the subsequent creation of a "second nature" as a technological totality that envelops human existence, he seems to assume there is a remnant of "nature" out there somewhere, and that technology can only replace nature from the perspective of the human. For example:

> Technique now constitutes a fabric of its own, replacing nature. Technique is the complex and complete milieu in which human beings must live, and in relation to which they must define themselves. It is a universal mediator, producing a generalized mediation, totalizing and aspiring to totality. (Ellul, "The Search" 23)

6 The problem is twofold: (1) market-driven eugenics cannot guarantee future beneficence; and (2) the emphasis on "hi-tech" means of enhancement, as opposed to socio-economic reforms or the equitable distribution of "old" technologies, is arbitrary. Langdon Winner nicely sums it up: "Better genes and electronic implants? Hell, what about potable water?" ("Are Humans" 44).

7

> The true principle of individuation is the genesis itself taking place, that is, the system in the course of becoming, while the energy actualizes itself. The true principle of individuation cannot be sought in what exists before individuation occurs, nor in what remains after individuation is accomplished; it is the energetic system that is individuating insofar as it realizes in itself that internal resonance of matter taking shape, and a mediation between orders of magnitude. The principle of individuation is the unique way in which the internal resonance of *this* matter is established in the process of taking *this* form. The principle of individuation is an operation. (Simondon, *L'individuation* 48; my translation, emphasis in the original)

8 At this point, the question of Simondonian politics is opened up. We don't have enough space to deal with it here. The reader is referred to the body of literature on the constitutive role of technologies in the shaping of transindividual dynamics. See Combes, Bardin, and Read.

9 Another reason is that, for Simondon, there is no difference between function and structure. A description of a function coincides with the performance of the relevant structures. Function is immanent to structure (see Vaccari, "El artefacto").

10

> [...] a phase is only a phase in relation to others, from which it distinguishes itself in a manner that is totally independent of the notions of genus and species. The existence of a plurality of phases finally defines the reality of a neutral center of equilibrium in relation to which there is a phase shift. This schema is very different from the dialectical schema, because it implies neither necessary succession, nor the intervention of negativity

as a motor of progress [...] (Simondon, *Mode of Existence* 173)

## bibliography


Akrich, Madeleine, and Bruno Latour. "Summary of a Convenient Vocabulary for the Semiotics of Human and Nonhuman Assemblies." *Shaping Technology/Building Society: Studies in Sociotechnical Change*. Ed. Wiebe E. Bijker and John Law. Cambridge, MA: MIT P, 1992. 259–64. Print.

Bailes, Jon. "The Political Theology of Neoliberalism." *State of Nature* 2018. Web. 10 Nov. 2018 <http://stateofnatureblog.com/adam-kotsko-political-theology-neoliberalism/>.

Barbrook, Richard, and Andy Cameron. "The Californian Ideology." *Science as Culture* 6.1 (1996): 44–72. Print.

Bardin, Andrea. *Epistemology and Political Philosophy in Gilbert Simondon: Individuation, Technics, Social Systems*. Dordrecht: Springer, 2015. Print.

Bostrom, Nick. "Human Genetic Enhancements: A Transhumanist Perspective." *The Journal of Value Inquiry* 37.4 (2003): 493–506. Print.

Bostrom, Nick. "In Defense of Posthuman Dignity." *Bioethics* 19.3 (2005): 202–14. Print.

Bostrom, Nick. "Why I Want to be a Posthuman When I Grow Up." *Medical Enhancement and Posthumanity*. Ed. Bert Gordijn and Ruth Chadwick. Dordrecht: Springer, 2008. 107–36. Print.

Bostrom, Nick, and Anders Sandberg. "The Wisdom of Nature: An Evolutionary Heuristic for Human Enhancement." *Human Enhancement*. Ed. Julian Savulescu and Nick Bostrom. New York: Oxford UP, 2009. 375–416. Print.

Cassirer, Ernst. *The Myth of the State*. New Haven: Yale UP, 1946. Print.

Cera, Agostino. "The Technocene or Technology as Neoenvironment." *Techné: Research in Philosophy and Technology* 21.2–3 (2017): 243–81. Print.

Clark, Andy. "Pressing the Flesh: A Tension in the Study of the Embodied, Embedded Mind?" *Philosophy and Phenomenological Research* 76.1 (2008): 37–59. Print.

Combes, Muriel. *Simondon: Une philosophie du transindividuel*. Paris: Éditions Dittmar, 2013. Print.

Ellul, Jacques. "The Search for Ethics in a Technicist Society." *Techné: Research in Philosophy and Technology* 9 (1983): 23–26. Print.

Ellul, Jacques. *The Technological Society*. New York: Vintage, 1964. Print.

Fisher, Mark. *Capitalist Realism: Is There No Alternative?* New York: Zero, 2009. Print.

Fuller, Steve, and Veronika Lipinska. *The Proactionary Imperative: A Foundation for Transhumanism*. London: Palgrave, 2014. Print.

Harari, Yuval Noah. *Homo Deus: A Brief History of Tomorrow*. London: Penguin Random, 2016. Print.

Harris, John. *Enhancing Evolution: The Ethical Case for Making Better People*. Princeton: Princeton UP, 2007. Print.

Hayles, N. Katherine. *How We Think: Digital Media and Contemporary Technogenesis*. Chicago: U of Chicago P, 2012. Print.

Heidegger, Martin. *Vorträge und Aufsätze*. Pfullingen: Neske, 1954. Print.

Hui, Yuk. "Cosmotechnics as Cosmopolitics." *E-flux Journal* 86 (Nov. 2017). Web. 3 Feb. 2018 <https://www.e-flux.com/journal/86/161887/cosmotechnics-as-cosmopolitics/>.

Hui, Yuk. "On Cosmotechnics: For a Renewed Relation Between Technology and Nature in the Anthropocene." *Techné: Research in Philosophy and Technology* 21.2–3 (2017): 319–41. Print.

Hui, Yuk. *The Question Concerning Technology in China: An Essay in Cosmotechnics*. Falmouth: Urbanomic, 2016. Print.

Hui, Yuk. *Recursivity and Contingency*. London: Rowman, 2019. Print.

Istvan, Zoltan. "Environmentalists are Wrong: Nature isn't Sacred and We Should Replace It." *The Transhumanist Wager* 13 Apr. 2019. Web. 14 Apr. 2019 <https://mavenroundtable.io/transhumanistwager/transhumanism/environmentalists-are-wrong-nature-isn-t-sacred-and-we-should-replace-it-TZ7Msb4mOk-B3n4kNqsyqg/>.

Ito, Joichi. "Resisting Reduction: A Manifesto." *JoDS* 14 Oct. 2017. doi:10.21428/8f7503e4. Web. 3 July 2019.

Kelly, Kevin. *What Technology Wants*. New York: Viking, 2010. Print.

Kurzweil, Ray. *The Age of Spiritual Machines: When Computers Exceed Human Intelligence*. New York: Viking, 1999. Print.

Kurzweil, Ray. "The Law of Accelerating Returns." *Alan Turing: Life and Legacy of a Great Thinker*. Ed C. Teuscher. Berlin: Springer, 2004. Print.

Kurzweil, Ray. *The Singularity is Near: When Humans Transcend Biology*. New York: Viking, 2005. Print.

Land, Nick. "Ideology, Intelligence, and Capital: An Interview with Nick Land." *Vast Abrupt* 15 Aug. 2018. Web. 4 Apr. 2019 <https://vastabrupt.com/2018/08/15/ideology-intelligence-and-capital-nick-land/>.

Land, Nick. "Meltdown." *Fanged Noumena: Collected Writings 1987–2007*. Falmouth: Urbanomic, 2011. Print.

Latour, Bruno. "On Technical Mediation: Philosophy, Sociology, Genealogy." *Common Knowledge* 3.2 (1994): 29–64. Print.

Latour, Bruno. *Pandora's Hope: Essays on the Reality of Science Studies*. Cambridge, MA: Harvard UP, 1999. Print.

Latour, Bruno. *Reassembling the Social: An Introduction to Actor-Network Theory*. New York: Oxford UP, 2005. Print.

Latour, Bruno. *Science in Action: How to Follow Scientists and Engineers Through Society*. Cambridge, MA: Harvard UP, 1987. Print.

Leroi-Gourhan, André. *Gesture and Speech*. Trans. from French by Anna Bostock Berger. Cambridge, MA: MIT P, 1993. Print.

Lyotard, Jean-François. *The Postmodern Condition: A Report on Knowledge*. Trans. Geoff Bennington and Brian Massumi. Minneapolis: U of Minnesota P, 1979. Print.

MacLeod, Ken. *The Cassini Division*. New York: Tor, 1998. Print.

Marx, Karl, and Friedrich Engels. *The Communist Manifesto: A Modern Edition*. London: Verso, 2012. Print.

Mayr, Otto. *Authority, Liberty and Automatic Machinery in Early Modern Europe*. Baltimore: Johns Hopkins UP, 1989. Print.

More, Max. "The Proactionary Principle: Optimizing Technological Outcomes." *The Transhumanist Reader: Classical and Contemporary Essays on the Science, Technology, and Philosophy of the Human Future*. Ed. Max More and Natasha Vita-More. Chichester: Wiley-Blackwell, 2013. 258–67. Print.

Morozov, Evgeny. *To Save Everything Click Here: The Folly of Technological Solutionism*. London: Allen Lane, 2013. Print.

Morton, Timothy. *Hyperobjects: Philosophy and Ecology After the End of the World*. Minneapolis: U of Minnesota P, 2013. Print.

Nitzke, Solvejg, and Nicolas Pethes. "Introduction: Visions of the 'Blue Marble': Technology, Philosophy, Fiction." *Imagining Earth: Concepts of Wholeness in Cultural Constructions of Our Home Planet*. Ed. Solvejg Nitzke and Nicolas Pethes. Bielefeld: Transcript, 2017. 7–21. Print.

Noys, Benjamin. "Drone Metaphysics." *Culture Machine* 16 (2015). Web. 15 Apr. 2017 <http://culturemachine.net/wp-content/uploads/2018/12/595-1453-1-PB.pdf>.

Read, Jason. *The Politics of Transindividuality*. Leiden: Brill, 2016. Print.

Rothschild, Michael. *Bionomics: Economy as Business Ecosystem*. Washington, DC: Beard, 2004. Print.

Sandrone, Darío. "Realismo tecnológico y diseño antropométrico: Dos ontologías técnicas." *La técnica en cuestión: Artificialidad, materialidad y ontología de lo creado*. Ed. Diego Lawler et al. Buenos Aires: Teseo, 2017. Print.

Simondon, Gilbert. *Imagination et invention (1965–1966)*. Chatou: Éditions de la Transparence, 2008. Print.

Simondon, Gilbert. *L'individuation à la lumière des notions de forme et d'information*. Grenoble: Millon, coll. Krisis, 1995. Print.

Simondon, Gilbert. *On the Mode of Existence of Technical Objects*. Trans. Cecile Malaspina and John Rogove. Minneapolis: U of Minnesota P, 2017. Print.

Simondon, Gilbert. *Sur la technique: 1953–1983*. Paris: Presses Universitaires de France, 2014. Print.

Sloterdijk, Peter. "The Anthropocene: A Process-State at the Edge of Geohistory?" *Art in the Anthropocene: Encounters among Aesthetics, Politics, Environments and Epistemologies*. Ed. Heather

Davis and Etienne Turpin. London: Open Humanities, 2015. Print.

Srnicek, Nick, and Alex Williams. *Inventing the Future: Postcapitalism and a World Without Work*. London: Verso, 2015. Print.

Stiegler, Bernard. *Technics and Time 1: The Fault of Epimetheus*. Trans. Richard Beardsworth and George Collins. Stanford: Stanford UP, 1998. Print.

Trawick, Paul, and Alf Hornborg. "Revisiting the Image of Limited Good: On Sustainability, Thermodynamics, and the Illusion of Creating Wealth." *Current Anthropology* 56.1 (2015): 1–27. Print.

Vaccari, Andrés. "El artefacto, ¿estructura intencional o sistema autónomo? La ontología de la función artefactual a la luz del intencionalismo, el dualismo y la filosofía de Gilbert Simondon." *Revista Iberoamericana de Ciencia, Tecnología y Sociedad* 19.7 (2011): 197–208. Print.

Vaccari, Andrés. "Why Should We Become Posthuman? The Beneficence Argument Questioned." *Journal of Medicine and Philosophy* 44.2 (2009): 192–219. Print.

Verbeek, Peter-Paul. "Materializing Morality: Design Ethics and Technological Mediation." *Science, Technology & Human Values* 31 (2006): 361–80. Print.

Verbeek, Peter-Paul. "Obstetric Ultrasound and the Technological Mediation of Morality: A Post Phenomenological Analysis." *Human Studies* 31 (2008): 11–26. Print.

Verbeek, Peter-Paul. *What Things Do: Philosophical Reflections on Technology, Agency and Design*. University Park: Pennsylvania State UP, 2005. Print.

Vinge, Vernor. "First Word." *Omni* Jan. 1983: 10. Print.

Winner, Langdon. "Are Humans Obsolete?" *The Hedgehog Review: Critical Reflections on Contemporary Culture* 4.3 (2002): 25–44. Print.

Winner, Langdon. "Decadencia y caída del tecnotriunfalismo." *REDES* 22.43 (2016): 127–42. Print.

yuk hui

# MACHINE AND ECOLOGY

## I introduction

> It might very well still take a considerable time to recognize that the "organism" and the "organic" present themselves as the mechanistic-*technological* "triumph" of modernity over the domain of growth, "nature." (Martin Heidegger, *Ponderings XII–XV*)

> [...] We lack creation. *We lack resistance to the present.* The creation of concepts in itself calls for a future form, for a new earth and people that do not yet exist. Europeanization does not constitute a becoming but merely the history of capitalism, which prevents the becoming of subjected peoples. (Deleuze and Guattari, *What is Philosophy?*)

> The river is the locality of the locale of the home. The river at the same time determines the becoming of human beings as historical in their being at home. The river is the wandering of that journey in which the becoming of being at home has its essence. (Martin Heidegger, *Hölderlin's Hymne "Der Ister"*)

This article titled "Machine and Ecology" hopes to investigate the relation between machine and ecology, and the philosophical and historical questions concealed in these two seemingly incompatible terms. First of all, I want to problematize these two ambiguous terms, "machine" and "ecology," as a preparation to de-familiarize and de-romanticize certain ideas about techno-ecology, and to suggest a *political ecology* of machines, which will center around what I term *technodiversity*. This quest for technodiversity belongs to a systematic inquiry of my thesis on *cosmotechnics* in *The Question Concerning Technology in China* (2016), which argues against certain traditions of philosophy, anthropology and history of technology, and suggests that instead of taking for granted an anthropologically universal concept of technics, one should conceive a multiplicity of technics, characterized by different dynamics between the cosmic, the moral and the technical.

Conventionally, one tends to think that machines and ecology are opposed to one another, because machines are artificial and mechanical while ecology is natural and organic. We may call this a dualism of critique (instead of a critique of dualism), since its mode of critique is based on the setting up of binaries, which it fails to go beyond, like the unhappy consciousness. This opposition has resulted from some stereotypes concerning the status of machines. Even today when people talk about machines they tend to think of mechanistic machines based on linear causality, for example, the digesting duck designed by the technician Jacques Vaucanson or the mechanical Turk by Wolfgang von Kempelen in the eighteenth century; and when they talk about ecology they tend to think of nature as a self-regulating system, which gives everything and takes back everything.

## 2 after the overcoming of dualism

The above-mentioned notions of machine and ecology undermine both the history of technology and the history of philosophy, therefore they also ignore the technical reality which conditions the validity of such a criticism. Criticism based on dualism fails to understand itself historically and critically. The mechanistic view of machines was already completely surpassed and rendered obsolete by cybernetics in the mid-twentieth century; instead, we have witnessed the emergence of a *mechano-organicism*. Today cybernetics becomes the modus operandi in machines ranging from smartphones to robotics and spacecraft. The rise of cybernetics was one of the major events in the twentieth century. Different from mechanism, which is based on linear causality, e.g., A–B–C, it rests on a circular causality, e.g., A–B–C–A′, meaning that it is reflective in the basic sense that it is able to determine itself in the form of a recursive structure. By recursion we mean a non-linear reflective movement which progressively moves towards its *telos*, be it predefined or auto-posited. Cybernetics belongs to a larger paradigm in the sciences, namely, organicism, which originated from the criticism against mechanism as a fundamental ontological understanding. Organicism also has to be distinguished from vitalism, which often relies on a mysterious (separate, immaterial) "vital force" to explain the existence of a living being; instead, it finds its foundation in mathematics. Cybernetics, as one form of organicism, mobilizes two key concepts, feedback and information, to analyze the behavior of *all* beings, both animate (living) and inanimate (lifeless), and both nature and society. In the first chapter of his *Cybernetics: Or Control and Communication in the Animal and the Machine* (1948), the founder of cybernetics, Norbert Wiener, first reiterated an opposition between Newtonian time and Bergsonian time. Newtonian motion is mechanistic, and time-symmetric, hence reversible, while Bergsonian time is organic, biological, creative and irreversible. It is not until the Second Law of Thermodynamics, proposed by the French physicist Sadi Carnot in 1824 (almost a century after Newton's death in 1727), that we recognize the "arrow of time" in being and the fact that the so-called entropy of a system increases with time and is irreversible. Already in his first book *Essai sur les données immédiates de la conscience* (1889), Bergson launched a fierce attack on the way in which time was conceptualized in Western science and philosophy. Time is here understood in terms of space, for example, in terms of intervals which can be represented in space. Therefore, the time that was thus conceptualized is actually timeless according to Bergson. It is also homogenuous, like the intervals marked on a clock. Instead, organic time or *durée*, Bergson suggested, cannot be fully understood as extension ordered in spatial terms, rather it contains heterogeneity or qualitative multiplicity in organic forms. Time is a force which is singular in every instant, like Heraclitus's river; it doesn't repeat itself twice like a mechanical clock. Indeed, mechanical or linear causality doesn't exist in correct duration. Bergsonian "organic" time also provides a new way to understand human consciousness and experience.

Wiener proposed that such opposition was already surpassed by the discovery of statistical mechanics in physics. For example, considering a container of particles, with statistical mechanics it is possible to communicate between the macrostate and the microstates, therefore, to control the behavior of the system. In other words, cybernetics endeavors to eliminate dualism, it wants to create a connection between different orders of magnitude, macro and micro, mind and body, like what Hans

Jonas says in his *Phenomenon of Life* regarding cybernetics, "an overcoming of the dualism which classical materials had left in possession by default: for the first time since Aristotelianism we would have a unified doctrine, or at least a unified conceptual scheme, for the representation of reality" (111). The same observation was made in Gilbert Simondon's *On the Mode of Existence of Technical Objects* (2012), where he considers the reflexive thinking of cybernetics (characterized by feedback and information) as key to the resolution of the dualism intrinsic in culture: traditional and modern, rural and urban, major (adult) and minor (child) modes of technology education, etc. In *Recursivity and Contingency* (2019), I put feedback under a more general category: recursivity. Recursion in general designates a non-linear operation which constantly returns to itself in order to know and determine itself. There are different modalities of recursions, but they all share the overcoming of dualism. Information is the measurement of the degree of organization, feedback is a recursive or circular causality which allows auto-regulation to take place. For example, when reaching my arm to grasp a bottle of water, many feedback processes are taking place, which allows me to adjust the attention of my eyes and the muscles of my arms until I reach the destination, or the *telos*. Therefore, towards the end of this chapter, Wiener was able to claim that:

> The modern automation exists in the same sort of Bergsonian time as the living organism, and here there is no reason in Bergson's considerations why the essential mode of functioning of the living organism should not be the same as that of the automation of this type [...] In fact, the whole mechanist–vitalist controversy has been relegated to the limbo of badly posed questions. (44)

Whether the claim of Norbert Wiener can be completely justified has to be scrutinized under the light of history. However, it remains significant for us to reconceptualize what is happening today regarding the relation between machine and organism, human and environment, technology and nature, departing from Wiener's cybernetics. Wiener's bold statement suggests a radical revaluation of the humanist values which oppose the organic and the inorganic, and it also renders the humanist critique ineffective. Different from what, for example, André Leroi-Gourhan and Bernard Stiegler might call "organized inorganic," Wiener's focus was not the man–machine or man–tool hybrid, but rather the possibility of assimilating both the organic and the inorganic by cybernetic machines. Modern machines are all cybernetic machines: they all employ circular causality as their principle of operation. In this sense, a cybernetic machine is no longer *merely* mechanistic, but rather assimilates certain behaviors of organisms. It is important to bear in mind that resemblance doesn't mean equivalence, and it is this misunderstanding that dominates our contemporary politics of machines today.

Ecology is equally a concept charged with ambiguity. If ecology is rooted in an attempt to understand the relation between the living being and its milieu, as it is in the case of Ernst Haeckel in the nineteenth century and is continued in the early twentieth century by Jakob von Uexküll, we have to bear in mind that this discourse remains important but insufficient to understand the complexity which belongs to human societies. Jacob von Uexküll has furthered Haeckel's concept of ecology to show that the environment is not only that which selects according to its physicality (in this respect Haeckel remains a Darwinian), but also that which is selected and internalized by the living being. The first type of selection may be called *adaptation*, meaning that the living being has to adapt itself to the milieu according to the available resources and physical conditions; the second type of selection may be called *adoption*, meaning that the living being has to select and to construct contexts from what is available to it as means of survival. The tick, an arachnid without eyes, remains inactive from its position on a tree, and only by detecting the smell of butyric acid (sweat), wind and warmth – which signify the approaching of a mammal – it falls down in order to attach to the animal's body, to reach the skin and then to suck its blood. There is a semiotics in the process of selection of information, based on the *Bauplan*, the sensorium and the central nervous system of the animal, which in turn defines its *Umwelt* (von Uexküll 50–51). However, human beings are not ticks, they invent tools and change the environment. They are beings talented not only with adapting to the external environment but also with changing and adopting that environment itself through technical means. In these processes of adaptation and adoption, we see that there is a reciprocity between the living being and its environment, which we can also call its organicity, namely, the fact that they do not only exchange information, energy and matter but also constitute a *community*. A human community is far beyond the sum of the human actors that constitute it, it also includes their environment and other non-human beings.

The intervention of human beings in the environment defines the process of hominization, the evolutionary and historical becoming-human and its politics. It is beyond our capacity to outline this process, however, human civilization could be seen as an intimate and complicit relation between humans and their environment, which gives rise to what is called *mesology* since Plato (according to Augustin Berque's historiography). However, to cut it short to the subject at hand, let us interrupt with a verdict from Marshall McLuhan, who said in an interview from 1974 that:

> Sputnik created a new environment for the planet. For the first time the natural world was completely enclosed in a man-made container. At the moment that the Earth went inside this new artifact, Nature ended and Ecology was born. "Ecological" thinking became inevitable as soon as the planet moved up into the status of a work of art. (49)

This verdict of McLuhan has to be further analyzed. The 1957 event, namely, the launch of Sputnik by the Soviet Union, is the first time that human beings were able to ponder upon the earth from the outside, and in this respect, the earth is now principally viewed as an artifact with the aid of space technology. In *The Human Condition*, Hanna Arendt also describes the 1957 launch of Sputnik as "second in importance to no other, not even to the splitting of the atom," because it suggests, like what Konstantin Tsiolkovsky said, that "mankind will not remain bound to the earth forever." This liberation from the earth directly confronts humankind with the infinite universe and prepares for a cosmic nihilism. It is the moment that nature ended and ecology was born. In contrast to the meaning that Ernst Haeckel gave to the term ecology towards the end of the nineteenth century meaning the totality of relations between a living being and its environment (2: 286–87; see also Richards 8), as well as to von Uexküll's definition of it as the selection process from the *Umgebung* (physical environment) to the *Umwelt* (the "interpretation" of the world by the living being), what McLuhan meant by ecology is no longer a biological concept. According to McLuhan, the earth is considered to be a cybernetic system monitored and governed by the machines upon it and in outer space. What we witness is the disappearance of the earth, since it is continuously absorbed into a plane of immanence constructed by the recursive thinking of cybernetics.

The hybridism between the natural environment and machines constitutes a gigantic system, and it is in this conceptualization that nature ended and ecology began. Ecology, beyond its strict use in biology,[1] is not a concept of nature but rather a concept of cybernetics. This is more evident when we refer to the notion of Gaia coined by James Lovelock to describe the ecological system of the earth: "a cybernetic system with homeostatic tendencies as detected by chemical anomalies in the Earth's atmosphere" (142). So we quickly arrived here at the position that the modern machine is no longer mechanistic and ecology is nothing natural; in fact, modern machines and ecology are two discourses adhering to the same principle, namely, cybernetics. The difference being, if we insist, that we have moved from individual machines, for example, those automatic machines in the factories of nineteenth-century Manchester described by Marx, to technical systems which connect different machines and establish recursivity between them. These systems can take different scales, from a local network to a planetary system such as the earth's technosphere. Now we want to ask what could be the implications of this redefinition of (the relation between) machine and ecology.

## 3 technological becoming of geophilosophy

We are more than ever in an epoch of cybernetics, since cybernetics was not a discipline parallel to other disciplines such as philosophy and psychology, but rather it aimed to be a universal discipline, able to unite all other disciplines, therefore, we could say, a universal (mode of) thinking par excellence. Cybernetics as a universal reflexive thinking has displaced the role formerly always played by philosophy as reflexive thinking. This displacement is not a rejection of philosophy, but rather, in the parlance of Martin Heidegger, the completion or end of philosophy (the German word *Ende* signifies both completion and end). What does this end mean? Does it mean that Western philosophy has no more role to play in the technological age, since it is already completed *in it* as its destiny? Or does it mean that philosophy in order to survive will have to reinvent itself, namely, to become a post-European (or post-metaphysical, post-ontological as you like) philosophy, and this is also no exception for Europe itself? I do not want to open a Pandora's box here, but I simply want to point out that cybernetic thinking as an alleged universal and ecological thinking is that which sublates, or at least pretends to sublate, the traditional metaphysical dualisms in ontology and epistemology, and it is in this respect that it calls forward a new condition of philosophizing, and therefore a new inquiry into the question of ecology.

Here is the postulation: maybe it is no longer a dualism which is the source of danger in our epoch, but rather a non-dualistic totalizing power present in modern technology, which ironically resonates with the anti-dualist ideology (e.g., rejection of any comparison between the East and the West). Ironically, because the anti-dualist ideology still believes that the main danger is dualism, without realizing that this duality is no longer the foundation of modern science and technology. In other words, without having examined this intimate relation between philosophy and technology, it will be difficult, if not impossible, to develop a philosophical thinking contemporary to our situation.

Now, let's bring our skepticism to the fore and pursue the argument further: will cybernetics be the solution to the ecological problems that we face today? Will the organismic model at the heart of cybernetics be able to overcome the shadow that European modernity has cast on us for centuries? If the early moderns provide us a mechanistic view of the world through geometrization (Kepler, Galileo, Newton, Descartes among others) and experimental science (Bacon and Boyles), now with cybernetics as the realization and concretization of organismic thought which started culminating since the end of the eighteenth century, *can we finally terminate modernity with cybernetics*? Don't we already find in cybernetics, and its planetary version, the Gaia theory, a generic logic that rests on the recognition of the relation between the living being and its milieu, that which the philosopher and orientalist Augustin Berque has emphasized in many places?

> To overcome the modern alternative is to recognize that the structural moment of our existence – our mediance – is such that each of us is split: "half" (in Latin, *medietas*, hence *mediance*) in one's individual animal body, while the other "half" consists of the eco-technical-symbolic system that is our life milieu. (Berque 60)

Berque has been endeavoring to propose a non-binary thinking that he found in Japanese thought or Eastern thought in general, and opposed it to the modern thought of which Cartesian dualism is the spokesperson. However, let us not rush to an answer because we may fall victim to the dualism of critique that we discussed above. Instead, let's consider the epigraph from Heidegger at the opening of this article concerning the relation between organism and technology. Heidegger saw that this becoming organic, or becoming ecological, is no more than the mechanistic-technological triumph of modernity over nature. This statement has to be accessed beyond the cynical impression that one may have at first glance. Heidegger's critique of cybernetics deserves our reflection today, since it is not about the celebration of the overcoming of dualism, but rather serves a call for prudence [*phronesis*] and a warning to avert from illusions and false analysis. Because at first glance, one may be able to claim that cybernetics has fulfilled an anti-dualistic critique of modernity. I would like to suggest, or even better to provoke, that with the rise of cybernetics and its organismic model, we may need a new agenda for mesology. We will have to understand this by rethinking the relation between technology and the environment. Instead of seeing technology as a result of the determination by the geographical milieu or that the natural milieu is destroyed by technology, one cannot neglect how the technology–environment complex constitutes its own genesis and autonomy, and how such genesis could be rethought or resituated in a cosmic reality which is proper to the milieu or *fûdo* (風土) in the sense of the Japanese philosopher Tetsurō Watsuji. We will elaborate on this point towards the end of the article.

To be brief – and this will deserve much more detailed analysis for sure in the future – this technological–environmental complex could be understood in two senses which are seemingly different yet remain intimately related. First of all, it is what the paleoanthropologist André Leroi-Gourhan terms technical milieu (*Milieu* 340–50). The technical milieu is that which acts as a membrane between the internal milieu conceived as an unstable and dynamic "mental tradition," and the external milieu consisting of the climate, natural resources and the influences of other tribal groups (334–35). Leroi-Gourhan uses the cell as an organic metaphor to explain the relation between these three terms and the permeability and resistance against technical tendencies. The technical milieu is that which is produced by the irreducible differences between the internal and external milieus, while at the same time it filters and diffuses what comes from the external milieu so that it can maintain the consistency of the internal milieu. In other words, the internal milieu and the external milieu form a reciprocal relation through the mediation of the technical milieu.

The second sense concerns a techno-geographical milieu, a term coined by Gilbert Simondon. It literally means that the geographical milieu including natural resources is no longer simply an object of exploitation but is rather integrated into the functioning of the technical object. In *On the Mode of Existence of Technical Objects*, Simondon gives us the famous example of the Guimbal turbine which successfully integrates the river as both the driving force (of an engine wrapped by oil under high pressure) and the cooling agent (57–58). In this case, the functionality of the river is multiplied, it becomes an organ belonging to the technical objects; the river is also what Simondon calls an associated milieu which provides a feedback mechanism for stabilizing and regulating the dynamic system: the stronger the current, the faster the turbine moves. Theoretically, more heat is produced which may burn the engine, while since the current is also fast, the heat can be more effectively carried away. The river and the turbine thus form a techno-environmental complex.

Both Leroi-Gourhan and Simondon were influenced by the metaphor of the organism in their conceptualization of the technical milieu and the associated milieu. This aspiration to an organismic or holistic model was a significant intellectual movement of their time. The role of the technical milieu in Leroi-Gourhan as a membrane between the internal and external milieus, was similar to what Simondon calls associated milieu, with the difference that Leroi-Gourhan still wants to single out the technical from the cultural (internal) and natural (external), while in the scheme of Simondon such distinctions have already disappeared. Simondon calls it a techno-geographical milieu (this is also the reason why Simondon was able to conceive of a conceptual plan for overcoming the antagonism between culture and nature, nature and technology, culture and technology). Simondon's interpretation of the significance of the Guimbal engine and the notion of the associated milieu was very much influenced by Wiener's cybernetics; and the reflective logic in cybernetics seems to Simondon to have displaced philosophy; and it is from this point that we can understand Heidegger's claim that cybernetics marks the end of philosophy. The river of Simondon stands in a peculiar relation to what Heidegger says in *The Question Concerning Technology* about the hydroelectric plant in the Rhine river, where the river becomes a mere standing reserve, to be constantly challenged and exploited by modern technology (16). Peculiar because at first glance, Simondon's formulation of the river as a techno-geographical milieu expresses an optimism, while Heidegger's description of the Rhine river as standing reserve is, though not necessarily pessimistic, a critique against the "technization" of *phusis*; however, they both refer to the same end of philosophy but with two different attitudes.

The emphasis of Simondon for what concerns the Guimbal engine is not simply about the exploitation of the river, but also it demonstrates a reciprocity between the technological and the natural, or what Simondon himself calls "co-naturality." The reciprocal and communal structure demonstrated by the Guimbal engine is only one case of the cybernetic thinking to which Simondon aspires for the overcoming of dualism or its more aggressive form, antagonism between culture and technology, nature and technology. After cybernetics, especially with the notion of "structural coupling" from biologists Hubertus Maturana and Francisco Varela, the technical functionality of the river described by Simondon seems to be present as a generic model of the techno-geographical complex. Environment is not only that which is *modified* by technology, but rather it is also more and more *constituted* by technology. Ecological thinking is not simply about protecting nature, but fundamentally a political thought based on environments and territories. Technology's increasing capacity to participate in the modulation of the environment forces us to develop a geophilosophy. This is by no means a new discovery, however, it is essential to analyze this historical trajectory in order to understand the stake of technological development today:

(1) The relation between the human and the environment is complexified in the course of time and the semiotics that defines perception and interpretation has to be constantly updated according to the evolution of technical objects in the sense of Simondon. The continuity and discontinuity from biological sensory detection, to the display of signs and symbols, and to the invention of electronic sensors that gradually cover the urban and rural area today entails a technological trajectory which constantly defines and redefines human and nature, which Peter Sloterdijk might term the domestication of human beings (89).
(2) The technology which is used for the domestication of livestock is fundamentally a modulation of the relation between livestock and its environment; or in other words, human beings intervene into the environment by controlling its fertility and sterility in order to modulate the behavior of the livestock on a large scale. Human communities maintain an apparent autonomy through the invention of laws, customs and symbolic systems which define taboos and transgressions. These constitute social norms and therefore also its opposite, social inadaptability, which is central to Michel Foucault's analysis.
(3) The technology of livestock domestication has gradually merged with the self-domestication of the human being, which may be understood in terms of what Foucault calls governmentality. Human beings' intervention into the environment constitutes a specific kind of governmentality which Foucault calls environmentality. At the beginning of this environmental thinking, we see that, and here I quote Foucault, "the population is the object that

government must take into account in all its observations and knowledge [*savoir*], in order to be able to govern effectively in a rational and conscious manner" (217).

(4) The control of population represents a *molar* type of governmentality which treats human beings in large quantities, therefore its technique can only be implemented through the mediation of laws and regulations which treat each subject as an equal and particular being. Technological inventions since the twentieth century supplement this molar mode of control with a *molecular* mode, meaning that each human being is treated as an individual, which differs from other individuals. Such an individual is defined by the relation between the individual and its environment constantly captured and capitalized in the form of data. This form of governmentality has become dominant during the coronavirus pandemic.

The generalization of recursive algorithms and its implementation in digital computers concretize cybernetic thinking and its applications in almost all social, economical and political domains. Capital moves from a mechanistic model accurately observed by Marx towards an organismic model realized by informational machines equipped with complex recursive algorithms. Data is the source of information, it is that which allows the recursive models to be ubiquitous and effective. The digital urbanism that is in the process of developing and which will be the central theme of the digital economy, is driven by the recursive operation of data. Data, in Latin, means something which is already given, like sense data that determines the falling of the tick or the red color of the apple in front of me; since the mid-twentieth century, data has acquired a new meaning, namely, computational information, which is no longer "given" from without, but is rather produced and modulated by human beings (Hui, *Digital Objects* 48). In this sense, we can see that the notion of "societies of control" described by Gilles Deleuze is far beyond the common discourse of a society of surveillance, it rather means societies whose governmentality is based on the auto-positing and auto-regulation of automatic systems. These systems vary in scale: it can be a global corporation like Google, a city like London, a nation state like China and also the whole planet.

## 4 ecology of machines

Here we would like to return to the question raised earlier: is cybernetics and its continuation in the twenty-first century via the systems theory of Niklas Luhmann et al. already a response to the critique of industrialism which inherited the dualistic tendency of early modern thought, like what Ludwig von Bertalanffy once outlined in his 1936 *General System Theory* that:

> The mechanical world view, taking the play of physical particles as ultimate reality found its expression in a civilization which glorifies physical technology that has led eventually to the catastrophes of our time. Possibly the model of the world as a great organization can help to reinforce the sense of reverence for the living which we have almost lost in the last sanguinary decades of human history? (49)

With the becoming reflective of cybernetic machines, is it possible to surpass modernity and hence the epistemologies which accompany it? Or is the generic model suggested by cybernetics for overcoming dualism still within the paradigm of modernity, as Heidegger suggested in the 1930s? What does it mean to be still within the paradigm of modernity? It means, I suggest, that it undermines the necessity of locality and diversity because it insists upon a universal episteme and upon the concept of progress.

Although it is true that machines are becoming organismic, it is in the permanent process of "becoming" as Simondon has observed, no matter how concrete a technical object is, that it still retains reminiscences of abstract schemes, while a living being is always already completely concrete. *It is within the parallax between the "not being completely concrete" and the illusion of being able to replace nature with digital informational technology that we find the question of politics today.* The former remains a humanist critique while the latter is rather transhumanist. Heidegger's response is neither humanist nor transhumanist, but rather, according to our interpretation, local. Being for Heidegger is a notion specific to a locality, which is called the land of the evening [*Abendland*]; not at least from a linguistic point of view, the concept of being didn't have its correspondence in Chinese language and thought (Graham 322–59). We could find it in his reading of Hölderlin's hymn "Der Ister," in which the river is conceived as both locality [*Ortschaft*] and wandering (or journeying) [*Wanderschaft*] at its origin (Heidegger, *Hölderlin's Hymn* 30). The river, which is the external milieu for Leroi-Gourhan and the associated milieu for Simondon, is the locality which is by default keeping stationary, and it is the wandering which moves forward. This seemingly contradictory movement, forward and backward, constitutes the historicity of "being there" (*Da-sein*). However, the destiny of locality is not yet clear in the technological epoch, and this ambiguity is the source of reactionary politics. It is because the question of the truth of Being can only take place in the danger brought about by human's frenzy for the gigantic, in the form of an event of appropriation initiated by a "shock of deep awe" (*Schrecken der Scheu*) (Heidegger, *Beiträge zur Philosophie* 8). However, shall one wait for this eschatology to happen, or should one instead take other paths which don't follow the history of Western thought as long as the universal has to be contested? It is the question of Being which Heidegger wants to explore that brings us back to the question of locality and historicity. The claim that with the invention of network technology, there is a compression of time and space sometimes prevents us from seeing what has always been already there and beyond. Indeed, one of the major failures of the twentieth century is the inability to articulate the relation between locality and technology and the reliance on an almost standardized ecological thinking endowed with a strong European humanism; technology became a provocation of either reactionary politics which is based on a dualism between tradition and modernity or a fanatic accelerationism which believes that the problems that we have and have inherited will finally be resolved by technological advancement, be it geoengineering for repairing the earth or the subversion of capitalism by accelerating towards full automation. From the economic and technocratic perspective, there is very little value in taking locality into consideration besides for its relevance to the availability of natural resources. The advancement of network technology will speed up the spatial compression and therefore there is no more value to discuss what could be called "geographicality" since all exchanges are done at the speed of light. This ignorance of the milieu is also an ignorance of locality; it fails to establish an intimate and complicit relation between the earth seen from the perspective of the territory and globalizing technology.

We still have to add why cybernetics is not yet sufficient as a non-dualistic solution before we arrive at understanding locality. The logic of cybernetics remains formal; therefore, it underestimates the milieu by reducing it to mere functionality based on feedback so that the milieu can be integrated into the operation of the technical object. In this respect, the milieu is exposed as a scientific and technological object, while its position within the genesis of technicity is ignored. This is also why until Part III of *On the Mode of Existence of Technical Objects*, Simondon declared that analysis of the evolution of technical objects and the analysis of the relation between the human and technics are not sufficient to understand technicity, and that it is necessary to *situate* technical concretization within the genesis of technicity, which means to relate technological thought to other thoughts. Simondon's unfinished project (judged from the standpoint of cosmotechnics) suggests conceiving a genesis starting with a magic phase which constantly bifurcates, firstly into technics and religion, then in the second stage, each of them bifurcates into the theoretical and practical part. Simondon understands technological development as a constant entanglement with religious thinking, aesthetic thinking and philosophical thinking oscillating between technology's exigency to diverge and thinking's necessity to converge. Technicity means here the cosmo-geographical specificity of technology and how such cosmo-geographical particularity has participated in shaping the technological mentality, which includes understanding of technology, a sensibility towards matter, form and other forms of existence, relation between art and spirit, etc. It is also for this reason that Simondon's project has to be pursued further by looking into the cosmological specificity of cultures. For example, Tetsurō Watsuji has tried to point out almost a century ago how the milieu affects the way of seeing and painting. The Japanese word *fûdo* comes from the two Chinese characters for wind (風) and soil (土). Watsuji classifies three types of *fûdo*, namely, monsoon, desert and meadow. To give brief examples of Watsuji's observations, he thinks that, since Asia is heavily affected by monsoons, the resulting relative lack of seasonal change creates an easy-going personality. In Southeast Asia especially, since the weather is always very warm, nature provides a plenitude of foodstuffs, and therefore there is no need to labor too much in order to survive, or to worry about the possibility of day-to-day living. Similarly, he argues that the lack of natural resources in the deserts of the Middle East creates solidarity between peoples, so that the Jewish people, although they live in diaspora, remain united; while in the meadowlands of Europe, clear and regular seasonal changes demonstrate the constancy of the laws of nature, thus suggesting the possibility of mastering nature with science.

This cosmological specificity gives rise to different technics, for example, in Greece the plenitude of sun and clear sky gives priority to the form, while the obscure *fûdo* in Asia gave rise to the style of haziness in painting (Watsuji 90). Cosmo-geographicality constitutes an important dimension of locality.

Cybernetic thinking remains a thinking of totalization, since it aims to absorb the other into itself, like Hegelian logic, which sees polarity not as oppositional but rather as a motivation towards synthesized identity. This is also why the Hegelian and cybernetician Gotthard Günther considers cybernetics fundamentally the operational (technical) realization of Hegelian reflexive – i.e., dialectical – logic (95). The complexification of cybernetic logic finally leads to an absolute totality. Bearing this in mind, without being able to reiterate Günther's interpretation of the place of Hegelian reflexive logic in cybernetics here,[2] we can probably formulate our claim as follows: *to think beyond cybernetics is to think beyond the totalizing effect of a non-dualist thinking*. In other words, how to reintroduce the question of locality into the discourse of machine and ecology today? And how does this reintroduction of locality contribute to the discourse on machines?

We are not opposing machine and ecology as if machines are those things that only rape Mother Nature and violate the harmony between the human and nature, an image attributed to technology since the end of the eighteenth century. We must advance instead towards an *ecology of machines*. Neither are we following the Gaia theory that the earth is a single super organism or a collectivity of organisms in line with the thought of James Lovelock and Lynn Margulis. Instead, I would like to propose to reflect on an ecology of machines. To open this ecology of machines, we will need first of all to go back to the concept of ecology. The fundament of ecology is diversities, since it is only with biodiversities (or multispecies including all forms of organisms including bacterial) that the ecological system can be conceptualized. To discuss the ecology of machines, we will need a different notion in parallel to *biodiversity*, which we call *technodiversity*. Biodiversity is the correlate of technodiversity, since without technodiversity, we will only witness the disappearing of species by a homogenous rationality. Take the example of pesticide, which is made to kill a certain species of insects regardless of their geographical locations precisely because it is based on chemical and biological analysis. However, we know that the use of the same pesticide may lead to different disastrous consequences in different environments. Before the invention of pesticide, there were different techniques employed to combat insects threatening the harvest of agricultural products, for example, natural resources found in the region. Namely, there was a technodiversity prior to the employment of pesticide as a universal solution. Pesticide is apparently more effective in the short term, but it is well established today that we have been only looking at our toes when we think about the distant future. We can say that technodiversity is fundamentally a question of locality. Locality doesn't necessarily mean ethnocentrism, nationalism or fascism, but rather, it is that which forces us to rethink the process of modernization and globalization and allows us to reflect on the possibility to *resituate* modern technologies. Locality is also key to conceive of a multiplicity of cosmotechnics. Locality doesn't mean identity politics here, but rather the capacity to reflect on the technological becoming of the local, not to retreat to traditionalism of one form or another, but rather for multiple localities to invent their own technological thought and future – an immunology or rather immunologies yet to be written.

What are the localities of non-European countries such as Japan, China and Brazil, today? Heidegger's long exposition on the relation between technology and Western philosophy is occidentally oriented. We should take the term orientation in a literal sense here, namely, as *Erörterung*, i.e., an identification as to where one is and as to what one will become. It is in this sense that Heidegger is also a thinker of geopolitics. To take up the project of Heidegger, but also to go beyond him today, is to carry his reflection beyond Europe. I want to put this challenge into a speculative question: for non-European cultures, can we identify their own technological thought in the same way that they also have different *fûdo*? Can these technological thoughts contribute to the imagination of technological futures, which are now unfortunately dominated by the transhumanist ideology? I tend to believe that it is possible and necessary to rediscover different technologies, which I call cosmotechnics. Cosmotechnics is not simply about, different ways of making things, for example, different techniques of knitting or dyeing. I gave it a preliminary definition in *The Question Concerning Technology in China* as the unification of the moral and the cosmic through technical activities. The term unification will have to be further elaborated,[3] for our purpose here, cosmotechnics should be understood as an *Urtechnik*, it challenges our current understanding of technology and therefore also its future. This cosmological specificity must be rethought beyond astral physics, beyond the conceptualization of the universe as a thermodynamic system; it also reopens the question of moral beyond ethical rules which are added posteriorly as constraints to new technologies. Technical activities unify the moral order and the cosmic order; and by unification I mean reciprocal processes which constantly enforce each other to acquire new meanings. This is the reason why I wanted to reinterpret what André Leroi-Gourhan calls *technical tendency* and *technical facts* (*L'homme* 27–35). Technical tendency is what seems to be universal, like laws of nature, for example, the use of flint to produce fire and the invention of wheels for transportation could be found in almost every civilization (we rarely see any triangular wheels if only in our imagination). Technical facts are the particular features that vary from one civilization to another; in the process of diffusion, technology was filtered and modified according to constraints intrinsic to the internal milieu. For Leroi-Gourhan, technical facts are determined by numerous factors, but largely by material constraints, while I tend to think that the differences in technical facts entails different cosmologies and their moral constraints, which encompass far more than functional aesthetics.

I would like to conclude by picking up the biochemist-turned-sinologist Joseph Needham's question here, namely, why didn't modern science and technology develop in China and

India but only in Europe? Historians who attempt to answer this question tend to carry out comparative studies on the advancement of technology in Europe and China as if the essence of technology concerns merely efficiency and mechanical causalities: for example, paper-making in the second century in China was more advanced than in Europe. However, this line of inquiry, it seems to me, has betrayed Needham's own stance. This is so because Needham was actually suggesting that there were two different trajectories of technology in China and in Europe, which were less constrained by material causes than by their different ways of thinking and forms of life. To put it in other words: answering Needham's question is not to show who is more advanced than the other, but rather to elaborate on the different systems of technological thought. It is the reason why *The Question Concerning Technology in China* was aiming to be a response to Needham by taking his implicit thesis further.

The technological upheaval since the nineteenth century has presented us with a convergence which at times seems inevitable while at the same time seems problematic and necessary to be *fragmented* in favor of other forms of convergence. The inquiry into the relation between machine and ecology is less about how to design more intelligent machines, but rather requires first of all a discovery of cosmotechnical diversity, while such diversity has to be thought through by going back to the question of locality, therefore re-articulating the concept of technics by resituating it within the geographical milieu, culture and thinking. The task that is left to all of us is the effort to rediscover these cosmotechnics in order to reframe modern technologies, namely, by reframing the enframing [*Gestell*]; only through such a reframing can we imagine a "new earth and people that do not yet exist" (Deleuze and Guattari 108).

## disclosure statement

No potential conflict of interest was reported by the author.

## notes

1 Note that a lot of biologists use this term and it is generally considered to be a biological discipline studying relationships of biotic and abiotic elements.

2 For a more detailed analysis please see Hui, *Recursivity and Contingency*, ch. 2.

3 I will elaborate on this notion of "unification" in my forthcoming book *Art and Cosmotechnics* (Minneapolis: U of Minnesota P, Nov. 2020).

## bibliography


Arendt, Hanna. *The Human Condition*. Chicago: Chicago UP, 1998. Print.

Bergson, Henri. *Time and Free Will: An Essay on the Immediate Data of Consciousness*. London: Allen, 1913. Print.

Berque, Augustin. *Thinking through Landscape*. Trans. Anne-Marie Feenberg-Dibon. London: Routledge, 2014. Print.

Bertalanffy, Ludwig von. *General System Theory*. New York: George Braziller, 2015. Print.

Deleuze, Gilles, and Félix Guattari. *What is Philosophy?* Trans. Hugh Tomlinson and Graham Burchell. New York: Columbia UP, 1994. Print.

Foucault, Michel. *Power: Essential Works of Foucault, 1954–1984*. Ed. James D. Faubion. Trans. Robert Hurley et al. New York: New, 2001. Print.

Graham, A.C. *Studies in Chinese Philosophy and Philosophical Literature*. New York: SUNY, 1990. Print.

Günther, Gotthard. *Das Bewußtsein der Maschinen Eine Metaphysik der Kybernetik*. Baden-Baden: Agis, 1963. Print.

Haeckel, Ernst. *Generelle Morphologie der Organismen*. Berlin: Georg Reimer, 1866. Print.

Heidegger, Martin. *Beiträge zur Philosophie (vom Ereigni)*. Frankfurt am Main: Vittorio Klostermann, 1989. Print.

Heidegger, Martin. *Hölderlin's Hymn "The Ister."* Trans. William McNeill and Julia Davis. Indianapolis: Indiana UP, 1996. Print.

Heidegger, Martin. *Ponderings XII–XV: Black Notebooks 1939–1941*. Trans. Richard Rojcewicz. Indianapolis: Indiana UP, 2017. Print.

Heidegger, Martin. *The Question Concerning Technology and Other Essays*. Trans. William Lovitt. New York: Garland, 1977. Print.

Hui, Yuk. *On the Existence of Digital Objects*. Minneapolis: U of Minnesota P, 2016. Print.

Hui, Yuk. *The Question Concerning Technology in China. An Essay in Cosmotechnics*. Falmouth: Urbanomic, 2016. Print.

Hui, Yuk. *Recursivity and Contingency*. London: Rowman, 2019. Print.

Jonas, Hans. *The Phenomenon of Life: Toward a Philosophical Biology*. Evanston: Northwestern UP, 2001. Print.

Leroi-Gourhan, André. *L'homme et la Matieère*. Paris: Albin Michel, 1973. Print.

Leroi-Gourhan, André. *Milieu et technique*. Paris: Albin Michel, 1973. Print.

Lovelock, James. *Gaia: A New Look at Life on Earth*. Oxford: Oxford UP, 2000. Print.

McLuhan, Marshall. "At the Moment of Sputnik the Planet Became a Global Theatre in which There Are No Spectators but Only Actors." *Journal of Communication* 24.1 (1974): 48–58. Print.

Needham, Joseph. *The Grand Titration: Science and Society in East and West*. London: Routledge, 2013. Print.

Richards, Robert J. *The Tragic Sense of Life: Ernst Haeckel and the Struggle over Evolutionary Thought*. Chicago: U of Chicago P, 2009. Print.

Simondon, Gilbert. *On the Mode of Existence of Technical Objects*. Trans. Cecile Malaspina and John Rogove. Minneapolis: Univocal, 2017. Print.

Sloterdijk, Peter. *Not Saved: Essays after Heidegger*. Trans. Ian Alexander Moore and Christopher Turner. Cambridge: Polity, 2017. Print.

Uexküll, Jakob von. *A Foray into the Worlds of Animals and Humans. With a Theory of Meaning*. Minneapolis: U of Minnesota P, 2010. Print.

Watsuji, Tetsurō. *Climate and Culture: A Philosophical Study*. Trans. Geoffrey Bownas. Westport, CT: Greenwood, 1961. Print.

Wiener, Norbert. *Cybernetics: Or Control and Communication in the Animal and the Machine*. Cambridge, MA: MIT P, 1985. Print.

bernard stiegler

*translated by daniel ross*

# NOODIVERSITY, TECHNODIVERSITY

## *elements of a new economic foundation based on a new foundation for theoretical computer science*

### I the "question concerning technology" and the ordeal of truth

Whoever we are, *we all* today face the "question concerning technology"[1] – and now (29 March 2020), we face this question in the experience of being confined by a pandemic. Confinement confronts us directly with our way of being simultaneously alone with our "connections" and collectively "connected," and the pandemic opens up in an immense way the question of the exosomatic factor – in the sense of what Alfred Lotka called "exosomatic evolution" (192) – in the pathology of the biosphere-cum-technosphere.

We should *very* carefully reread Georges Canguilhem's *The Normal and the Pathological* from this standpoint, through which we may finally begin to realize that "biopolitics" is above all a technopolitics, that is, a politics of noetic life in so far as it forms, with animal and plant life, but also with bacteria and viruses, a very singular archipelago, which with the Digital Studies Network we have called the archipelago of the living.

Before the pandemic, all of us were more or less in the midst of an encounter with the *Frage nach der Technik*, due to the consequences of one fact: computational technology has overturned our ways of life – by being *concretized* in myriads of "apps," services, databases, in software and in algorithms, centrally controlled by a few platforms based mainly in two countries.

This concretization (in the Simondonian sense) led an associated techno-geographical milieu (Simondon, *On the Mode* 57–58; Stiegler, *Automatic Society* 22) to be established in which the *element* is no longer, as in the case of the Guimbal turbine studied by Simondon, tidal waters, but rather the "human resource" provided by individuals, who have become "inforgs" (Floridi), and who, in so doing, individually and collectively constitute the functions of a reticulated planetary exorganism that is extraordinarily vulnerable and dangerously dependent.

In the future, computational technology is likely to bring yet further changes to ways of living (and not just to lifestyles). A reconsideration of its functions, shortcomings, limits and dangers must be placed at the heart of the debate concerning what we want the nature and culture of the post-pandemic world to be, but also and equally, the post-data-economy world, given that the data economy, based on an exploitation of machines and through them of the people who serve them, is skewed towards the abyss into which it is rushing: entropy.

## 2 virology and virulence: the new dispute

This change over the last twenty-seven years brought by generalized reticulation (which in February 2019 involved 4.4 billion earthlings, according to Amsili and Maussion) mostly presents itself as a series of increasingly inextricable problems, to which are now added the problems of virology – and therefore of virulence, from the Latin *virulentus*, which initially meant "venomous, poisonous."

It is from this point of view that we should read Jean-Luc Nancy's response to his friend Giorgio Agamben, when the latter accused states of using "something not too different from the normal flus" as a pretext for perpetuating a state of exception:

> the exception is indeed becoming the rule in a world where technical interconnections of all kinds (displacements, transfers of all kinds, absorption or diffusion of substances, and so on) are reaching a hitherto unknown intensity that is growing along with the population. In rich countries, too, this increase in population entails a longer life expectancy, and thus an increase in the number of elderly people and, in general, of people at risk.
>
> We must be careful not to hit the wrong target: an entire civilization is in question, there is no doubt about it. There is a sort of viral exception – biological, informational, cultural – that for us becomes pandemic. Governments are nothing more than grim implementers, and attacking them seems more like a diversionary manoeuvre than a political reflection. (Nancy; translation modified)

If there is indeed a state of exception, Nancy replies to Agamben, it is "a sort of viral exception." These lines contain the expression of an *ordeal of truth* with regard to the *question concerning technology* – and with regard to the relationship between technics and life.

I would like to show here that:

- the challenge for thinking, especially after the pandemic, is and will always be to transform the *problems* posed by technics – become technology – into *questions*, that is, to turn technics into the object par excellence of thinking;
- the question of the exception is indeed essential, at once central and peripheral, and must be considered not just with Carl Schmitt, Walter Benjamin, Martin Heidegger and Michel Foucault, but with Vladimir Vernadsky, Georges Canguilhem and Alfred Lotka, and as the question of the relationships between entropy, negentropy and what Giuseppe Longo and Maël Montévil call anti-entropy;
- in the case of the noetic form of life, the question of anti-entropy must itself be transformed, from the exosomatic standpoint developed by Lotka, into that of anti-anthropy, by reconsidering the human fact from the perspective of a neganthropology;
- ultimately, and as a new foundation of political economy on a technospheric scale, as well as a critique conducted *from* this scale, which is also cosmotechnical, these questions, which are both conceptual and ideal, must be explored via a re-evaluation of the role of computer science [*informatique*] and cognitivism in the neoliberal apparatus, which has now become ultraliberal and libertarian; these questions require, from the basic sciences as well as from philosophy, economics and law, a new foundation for theoretical computer science – in the service of an anti-anthropic and neganthropological conception of the functions of machinic and automated calculation in the reticulated societies of the technospheric era.

This is why I am not sure that I fully share Jean-Luc Nancy's point of view when he posits without further specification that, in relation to what he calls the "viral exception," governments are "nothing more than grim implementers, and attacking them seems more like a diversionary manoeuvre than a political reflection." If it is obviously not a question of "attacking" anyone (as Nietzsche warned us long ago), we must nevertheless be aware that this kind of crisis of *pharmacological virulence*, and in that way of *virology in exosomatization*, will certainly not spare us from a regression precipitating the potential (murderous) "evil" of resentment towards scapegoats, which we can struggle against only by forging the weapons of a new critique (of scientific power, technological power, economic power and political power – and therefore of governments, and those who serve them, directly or indirectly).

## 3 changing *pansements* to re-found theoretical computer science

What is at stake here is precisely the struggle against this regressive tendency, where:

- it is *firstly* a question of knowing the extent to which such a tendency can affect and infect a thought that claims to distance itself from this type of virus[2] – a question that is part of what I have called the *careful thought [pansée] of noetic bandages [pansements] that always end up themselves getting infected* (Stiegler, *Qu'appelle-t-on panser? 1*);
- it is *at the same time* a question of developing a new political reflection that *challenges* and *calls into question* (like the noetic treatment or bandage [*pansement*] that any form of philosophical thinking must always be) these governments, both upstream and downstream of these problems, and that challenges them and calls them into question – along with the economic powers that support and manipulate them – *along with all those who have been in solidarity with them*, from near or far, directly or indirectly, *both positively and negatively*.

Being in "negative solidarity" means participating in the role-playing of good guys and bad guys, who are "as thick as thieves" in that all of them want to be part of the same *show* – so as to change nothing, neither the cast nor the script, instead merely *rearranging the sets and the staging*. It is this spectacle that generates what are called "postures."

Being *compromised* in this way – which it is not a matter of seeing as the corruption of the bad guys, but rather as a particularly thorny problem of noetic pharmacology, and where this is a problem that *we all negatively and intermittently fall into*[3] – is something that concerns *everyone*, but *more particularly* concerns those professional thinkers that academics are in principle.[4]

In principle, academics *bear* in principle (that is, at the *scale* of a principle) the universal, and, in the name of this principle – namely, that of a *profession*, which, as a profession of professors, is necessarily always a "profession of faith" – they contribute (in principle – from the *arkhē*) to society (of which "governments" are a key dimension: that of decision) being *gripped* by what, through the problems that a society and a government encounter – including *the problem of a universal that always tends to become hollow, and thus to serve as an argument against the diversal*, which is also the local or peripheral (which is also to say, the plebeian, the Gilet Jaunes, migrants, and so on) – this professorial and academic body *establishes* as questions.

The problems that this society encounters are first of all the problems that it provokes, as a whole and through its contradictory behaviours, which are also dynamic contradictions – of positions and interests, that is, classes – where academic thinkers, who are in this respect "functionaries of humanity," must be able to *constitute* these problems as established questions (and here, where the question of the *established* arises, it is not at all a question of restoring a figure of "mastery" constitutive of "modern metaphysics," as every heir of Jacques Derrida *tends* to believe spontaneously with an atavistic and parochial attitude – that is: they tend to do what in French is called "hiding behind one's little finger").

Theoretical computer science has been *abandoned* by care-ful thinking [*pensée qui panse*], and more particularly by European philosophy,

"French theory," the heirs of Marxist thought and psychoanalysis, with the exception of Félix Guattari. That this specific dimension of our time has been abandoned to the ideologues of neoliberalism, who hide behind their computational pseudosciences based on a confusion of science and quantification (and which are cognitivist in this sense), is due, at the same time, to the fact that:

- philosophy no longer *practises* mathematics;
- mathematics has tended to merge with mathematical physics, and through that with an obsolete yet hegemonic *mechanism*, something that Husserl already identified in *The Crisis of European Sciences* as a tendency dating back to algebraization;
- these forms of mathematics inspired by mathematical physics and applied via algorithms have been appropriated by a rudimentary and even destitute cybernetics, by an information theory that is merely a *bricolage*, and by a use of technics that consists above all in eliminating any thought and all (noetic) *pansements* for such a technics.

This amounts, then, to a way of taking up the *Frage nach der Technik* – and this is what I will attempt to do in what follows, as a way of opening up a dialogue with Yuk Hui, and, through him, between what used to be called the West (which now consists only of ruins under a setting sun) and what is still and more than ever called China.

## 4 the entropic orientation of the capitalist stage of exosomatization and the failure of philosophical engineering

In the conceptual framework that I have tried to develop over the past decade, the problems caused by technology raise questions about the *pharmacology of exosomatic organs* at different scales (*shifts* of scale, based on *relations* of scale, presuppose distinctions between *levels*, which constitute the various scales of locality ranging from the cell to the biosphere, and to the exosphere that now surrounds it, constituting the technosphere through this *ring* by which life and humanity find themselves encircled – like the iris and the pupil of a tired cyclopian eye).

The change in our ways of living, effected through the generalized digitalization of existences that today occurs in *all* dimensions (and this *totality* itself raises an immense and immensely new question), this change, whether already accomplished or yet to come, seems to be skewed, *one-sidedly oriented*, in a sense that each day seems increasingly inevitable and damaging: this one-sidedness is that of entropy.

Digital change, which has now been concretized as a process of *planetary reticulation* (see Longo) *dominated by network effects*, and therefore herd effects, and hence by virality (as "memes"[5] and "mimetic desire"[6]), proves day after day to be ever more disappointing, and each day contains less of the hopes that had been aroused – over the course of about two decades (roughly between 1985 and 2005).

These hopes, which now appear instead to be lost illusions, first emerged in the restricted circle of free software designers – an organization of developmental work (developing coding) based on sharing the knowledge formed through these activities, completely different from the classical model of the industrial division of labour (and on the basis of a new intellectual property right), with principles forged around 1985 at Berkeley and MIT. (Only André Gorz really saw the fundamentally new character of this organization of industrial work, after two centuries of the hegemony of the industrial division of labour as analysed by Smith, then Marx, then Durkheim.)

Ten years after the interventions of Richard Stallman led to the Free Software Foundation, which will bear the seeds of a philosophy of right in some way based on "philosophical engineering" (an expression notably used by Tim Berners-Lee[7]), these hopes became much more widely shared with the public launch of the World Wide Web – which would also generate the "tech bubble" and the speculative "madness" of Nasdaq.

All of this, however, will turn out to be the *preliminary* basis for a *movement to automated financialization*, initially launched in the early 1970s with the aid of Bernard Madoff,[8] which undoubtedly originally contributed to the process of financialization that itself paved the way for the new wave that would be known as "neoliberalism" and the "conservative revolution."

These "hopes" of escaping from, on the one hand, the condition (in the Arendtian sense) of a "one-dimensional man" resulting from the culture industries and from the age of *Dummheit* specific to what Adorno and Horkheimer called barbarism (xiv), and, on the other hand, from the establishment of a cybernetico-nuclear age that appropriates the potential contained in the decentralization of networks and the editorialization of the World Wide Web based on feedback loops and a computational regime of recursivity, and of what I more generally call recurrence[9] – these hopes would begin to decline, and in hindsight to become dangerous illusions, when social networks, combined with smartphones, began to put an end to the social web, also called the Web 2.0, and more generally to the logic of the web itself.

Let us make clear, here, that the Ars Industrialis association was set up above all by positing the *inherently pharmacological* character of these technologies, understood also as new stages of the process of grammatization, after those that struck the West (and struck its coins and founded its markets), on the one hand as linear writing, and on the other hand as mechanical grammatization (making possible automation). Let us also note that it was by combining these achievements with the Chinese technologies of printing, maritime navigation and explosives – making possible primitive accumulation – that the modern West was able to impose its global domination on a world that was on the way to becoming a technosphere (and is now in the process of de-Westernization).

The "end" of the "social web" began with the appearance of smartphones and social networks and was subsequently combined with the 2008 financial crisis, which led to an intense acceleration of "platformization" (Daniel Ross, "Carbon and Silicon," in Stiegler and the Internation Collective, ch. 10). Becoming thus disillusioned, these hopes were then *massively instrumentalized* by the libertarian/transhumanist movement – thus continuing the mutation of neoliberalism after the Thatcher–Reaganite conservative revolution that defined the end of the twentieth century, a mutation concretized through the process of "smartification" described by Evgeny Morozov.

## 5 dystopias and improbabilities

More or less consciously – and more or less overwhelmed – we all believe that this becoming, which seems bound to become more "dystopian" every day, *unless something utterly improbable happens*, will continue *ever more systemically, and therefore more inevitably: until the break* [*rupture*]. And we all wonder whether this viral archi-event will not in one way or another turn out to *be* this rupture – which would not, obviously, necessarily be good news, unless it can be turned into a massive reinvention.

The dystopian character of the current state of affairs with respect to digitalization is due to what Geert Lovink calls "platform nihilism" – and to what we at Ars Industrialis have called "net blues" (Stiegler, *Automatic Society* 13). As for the meaning of "break" with regard to the pandemic, we can all understand that it is an opportunity to renew the "shock doctrine" (Klein), foreshadowing a new leap towards the *totalizing radicalization* of the reticular and technospheric Leviathan. What precisely would this doctrine be, if it exists, and what would be a counter-doctrine?

According to my own earlier analyses, we should *firstly* ask here: what does *improbable* mean in the era of the hegemony of probability calculations? The improbable is diversity – in this case, biodiversity, noodiversity. The probable (which is always *relative to the most probable*[10]) is the entropic tendency towards the elimination of the diverse. But the improbable is also the unexpected, the unforeseen, the "black swan," which is also to say, the *flaw* at the origin of a possible collapse.

The feeling of being overwhelmed derives above all from this obvious fact: the world in which we live is based on calculation, and calculation has itself become hegemonic with informational machines – such that information is functionally calculable – where the hegemony of calculation achieved by conceiving information in this way (as the result of a calculation) amounts to the *elimination of the diverse*, that is, of the beneficial improbable that alone is capable of *deferring* the multiplication of unforeseen, toxic improbable events.

Calculation, which cannot calculate *everything*, contrary to what the operators of "soft totalitarianism" strive to impose through their totalizations, generates that unforeseen against which only diverse improbabilities *protected from calculation* could prove resilient. This is the starting point of a proposal to *relaunch research into theoretical computer science* – whose central argument, however, will not be presented here, though it is based on analyses partly summarized in a text entitled "Capitalism as Epistēmē and Entropocene" (Stiegler, *The Neganthropocene* 139–51), of which I here recall three points:

(1) Capitalism is an *epistēmē* materialized by the fixed capital of the reticulated production apparatus that capital has become. This *epistēmē* hegemonically reconfigures *all instruments of calculation* by functionally integrating[11] them as instruments of statistics, measurement, simulation, modelling, observation, production, logistics, mobility, guidance, bibliometrics, scientometrics, marketing, self-quantification (the "quantified self"), and so on.

(2) Information is the allagmatic (see Simondon, *L'individuation* 529–36) operator of this *epistēmē*, a computational technology utterly homogeneous with a capitalism that subjects all those exchanges in which psychic and social life consists to market calculations. This calculation, which sets up *reticulated artificial intelligence*, is based on *cognitivism* as the general paradigm of all forms of knowledge.

(3) The cognitivist *epistēmē*, however, is an anti-*epistēmē*: it can develop only by installing a process of *generalized proletarianization*.[12] The correlationist mythology of "big data," as it is developed, for example, by Chris Anderson in "The End of Theory,"[13] is a perfect example of the way in which *ideology* has been reconfigured – both through the cognitivist paradigm and through marketing, which is itself now reticulated, mimetic and computational.[14]

In the remainder, we will limit ourselves to investigating the question of the *relationships between improbabilities and forms of exosomatization*, in particular with respect to standardization, and in so doing attempt to enter into a dialogue with Yuk Hui's thesis concerning what he calls, on the one hand, technodiversity, and, on the other hand, and correspondingly, cosmotechnics.

## 6 the dynamic supports of the "third world" and the present dissolution of reason

We feel and see that a *systemic* elimination of diversity is underway, and that it has everything to do, on the one hand, with *technology*, and, on the other hand, with *calculability* – technology being rationalized and through that inherently tied to calculation, which is not the case for technics, which is empirical, and inherently tied to know-how [*savoir-faire*]. This elimination makes us suffer, but we are not aware of it, we do not *know* it, and for a very specific reason: the knowledge that would allow us to think and care [*de penser et de panser*] for this fact, and thus to bring it to the stage of law, has still not been developed. Its constitution will be the key issue both in the aftermath of the pandemic and in the re-foundation of a theoretical computer science.

The project of relaunching a theoretical computer science that would take the need for diversity functionally into account presupposes the constitution of a neganthropology – so as to effect a bifurcation in the Anthropocene era and towards the Neganthropocene (Stiegler, *The Neganthropocene*). Such a neganthropology presupposes digital studies, whose principles and primary objects were proposed at the Centre Pompidou in 2012. It has since given rise to an informal network, the Digital Studies Network, and to a book, *Digital Studies: Organologie des savoirs et technologies de la connaissance.*

It was argued in *Automatic Society* and in the French reissue of *La technique et le temps* (in a new afterword entitled "Le nouveau conflit des facultés et des fonctions"[15]) that *automatic calculability involves a delegation of the analytical function of the understanding to an automatic retentional system* that leads to a *hypertrophy of the understanding* and to a *regression of reason* in the Kantian sense – as the faculty of deciding, operating through a synthesis that is also called judgement. In his post-thermodynamic resumption of these Kantian questions, Alfred Whitehead revived this singularity of a synthetic function that *would not be soluble into the analytical faculty.* And in *Objective Knowledge*, Karl Popper highlighted the irreducible link between what he called the "third world" and exosomatization.

This position on the contemporary hypertrophy of the understanding and on the function of reason follows on from the argument set out in *Technics and Time, 3*, where it was posited that the imagination, as Kant defined it in the first edition of the *Critique of Pure Reason*, then the schematism, as he explains it in the second edition, are not transcendental dimensions of the mind, but configurations that are established and metastabilized through hypomnesic tertiary retentions (the reissue of *La technique et le temps* specifying that these first appear in the Upper Palaeolithic). All of this relates to Popper's "third world": hypomnesic tertiary retentions are the dynamic supports of this third world.[16]

The possibility of *dissolving the singularity of reason via the automated understanding* has been the basis of *computational cognitivism*, which in recent decades has been imposed on a vast scale, together with so-called analytical philosophy and contemporaneous with the spread of digitalization. This domination will prove to be utterly homogeneous with neoliberalism: Friedrich Hayek and Herbert Simon met during the restoration of the so-called "neoliberal" doctrine undertaken by the Mont Pelerin Society (with which Popper was also involved), positing that the *market* is a *system of information*, and that *everything that is good is calculable* in terms of such information.

Both Hayek and Simon, then, prescribed the notion of democracy as a space for free information – but where such freedom actually means the freedom to reduce all reality to calculability, that is, to *subject* all *realization* (all future) to the hegemonic law of the market and all knowledge to this entropic condition. Unfortunately, Popper compromised himself in this coup against real democracy, which is always the protection, not of the majority, nor for that matter of the minority (these are accounting concepts), but of *diversity* – which must be cultivated in each citizen, and as a potential for neganthropic and anti-anthropic resilience.

## 7 why *precisely* is the question of technodiversity so *essential* at the end of the anthropocene era?

When Yuk Hui raises *the question concerning technology in China*, and initially presents this question of technics (before entering into the question of technics *in China*) as that of *technodiversity*, he takes this notion of diversity as a way of challenging the hegemony of a universal calculability that would be concretized by a machine that prescribes arrangements, themselves calculable, between:

- on the one hand, this machine and the *technical system* within which it functions, of which it becomes at the same time the clock, the memory and the central unit, that is, the operator (as *processing* unit);
- and, on the other hand, social systems, along with biological and geographical systems, where this thoroughly computational technical system *takes hold of all these systems by calculating them*, and where, in so doing, it *dissolves* them through *feedback loops* – themselves based on *recursive functions*,

> the difficulty of which was emphasized by Pierre Livet – that operate continuously, in real time, and by *reducing every transaction to a calculation on a market.*

Note here that:

- The *computer*, wrongly defined by cognitivists as a Turing machine, becomes the *cellular* element (in the sense that it becomes a smart-cell-phone) of a computational network collecting, processing and distributing data through feedback loops working several million times faster than the nervous systems that form those local networks that are also called, depending on the context, psychic individuals, citizens or consumers.
- As the aggregation of *psycho-technical apparatuses* that Luciano Floridi calls inforgs, the technosphere amounts to a megamachine in Lewis Mumford's sense, and, in this way, it constitutes a new type of higher complex exorganism (Stiegler, *The Neganthropocene* 132–34; *Nanjing Lectures* 286), where the platforms that exploit this megamachine claim to replace *political sovereignty defined by its aims* with a *sovereignty of computational efficiency that utterly ignores formal causes, final causes and material causes.*
- It is in this context that today the question of technodiversity imposes itself on all "global citizens," while from Yuk Hui's perspective it arises firstly from the perspective of China inasmuch as the latter is not reducible to the West – of which, moreover, it has become the great challenger.

It is in this contemporary context that it is appropriate, for example, to *return to the conditions in which Leibniz paid close attention to Chinese writing* – and to what it contains that may present a sovereign dimension of the spirit that would perhaps have always escaped the Western mind.[17]

The dissolution of reason is now effected by the "data economy" as an industry of traces, but it got underway in the twentieth century with the culture industries that were already substituting *rationalization* for *reason* (*modernity* being achieved at this price, as *modernization*). This literally disintegrating and hyper-rationalized dissolution (*logos* being entirely replaced by the *ratio* that is the algorithm, that is, *the* ratio – and every ratio, in the accounting sense, is, as soon as it appears, an algorithm) is equally a disintegration of the very notion of universality. And this is accomplished today through a misuse of the notion of the machine, and in this case the abstract machine theorized by Alan Turing.

This is how Western universalism can become, not an emancipatory reason that respects the diversity of people, but a rationalization that alienates all resources according to its own (that is, Western) interests. This process imposed by cognitive technologies[18] continues and completes what was initiated with cultural technologies[19] on a completely different register than spiritual technologies.[20]

Here, therefore, the question arises of knowing if *modernity* means saying yes or no to the universality of those tendencies historically developed by the West, and especially as technology, which concretizes the laws of universal science – and in particular, on the one hand, the physics of Newton, and, on the other hand, the mathematics of calculability. To approach these questions, we must first return to André Leroi-Gourhan's notion of universal technical tendencies, the characteristic features of which are briefly recapitulated by Yuk Hui in *The Question Concerning Technology in China* (8–10).

These questions become unavoidable at the end of the Anthropocene (as announced, variously, by the IPCC (Intergovernmental Panel on Climate Change), IPBES (Intergovernmental Science-Policy Platform on Biodiversity and Ecosystem Services), the signatories of "Approaching a State Shift in Earth's Biosphere" (Barnosky et al.), the appeal made on 13 November 2018 in *BioScience* (Ripple et al.), the February 2020 "L'appel de 1000 scientifiques" in which a thousand French scientists call for rebellion, and so on), because this era makes obvious the *extreme danger brought by the elimination of biological and noetic diversity*, where the operator of such a liquidation is the current stage of exosomatization that functions as generalized digital grammatization.

At this stage, which presents itself in a more or less eschatological way (in a sense that is not here religious), how should we interpret what Leroi-Gourhan calls universal technical tendencies?

## 8 tendencies and milieus

Should we, for example, follow the point of view of the accelerationists (taken up by Toni Negri), according to which it would be a matter of re-orienting reticular platform technology in the right direction (that of social justice and a re-established economic rationality, if we follow Nick Srnicek and Alex Williams, for example)? Or should we *diversify* techno-industrial concepts – a bit like the aftermath of the crash of 1987, when it was asserted by the American stock market authorities that an "idiomatization" of automated trading programs was essential in order to avoid systemically reinforcing downward trends (see Distler), as occurred, according to studies carried out afterwards, when then Secretary of the Treasury James Baker publicly suggested that prices were too high?

With respect to technodiversity, and in order to understand Yuk Hui's position with respect to it – which seems to be an objection to what Leroi-Gourhan says, and consequently to my own work in so far as I take up his thesis of universal technical tendencies[21] – two points must be taken into consideration:

(1) The question of the milieu, as Leroi-Gourhan poses it in *Milieu et techniques*, and as the milieu of the "ethnic cell" (here, Leroi-Gourhan evokes by analogy the work of Claude Bernard) is always already divided into *mutually diffracting* milieus, in the form of the *interior* milieu, the *technical* milieu and the *exterior* milieu – in the process of exosomatization[22] that Leroi-Gourhan himself called exteriorization; in this way, there is never *the* milieu (no more than there is *the* language), but milieu*s*, which are *originarily divided* (which can and must be compared to the fact there are always already language*s* – idioms, idiolects, dialects, patois) in a more or less noticeable way.

(2) On the other hand, we must differentiate exo*somatic* exorganogenesis and exo*mnesic* exorganogenesis: exomemorization (see Stiegler, *Qu'appelle-t-on panser? 2*) as the production of *hypomnesic supports*; and, starting from these *differently différant differentiations*, so to speak, we must make out the conditions of the *expression* of universal technical tendencies, which are *only* tendencies, and which, when they are *fully* realized, are never realized other than *as nightmares*: by destroying everything that in principle makes them diffract neganthropically, a diffraction raised in principle by Leroi-Gourhan but ignored by Simondon when he postulates a process of machinic concretization – leading to the constitution of associated milieus whose reticulated expansion (via cybernetics) constitutes what Heidegger called *Gestell*.

Before moving on to further analysis, a commentary on these two points is required.

## 9 what is a tendency?

The conditions of the *expression* of universal technical tendencies – that is, of the limitation and orientation of the process of technical individuation, the expression of which is always limited – vary according to the types of exomnesic grammatization through which these expressions are *produced*.

Since the Upper Palaeolithic (during which the process of grammatization began), exomnesic grammatization has produced the hypomnesic supports of social systems (in the sense of Gille and Luhmann: from shamans to academies, research bodies, courts and jurisdictions, parliaments, and so on, via churches, temples and theocratic organs of all kinds). The conditions of the *expression* of universal technical tendencies vary precisely according to the types of advances made in the processes of exomemorization – tied as they are to grammatization.

What is it that makes tendencies *tendencies*, and *only* tendencies? It is firstly because *a* tendency does not exist: it always plays out along with a counter-tendency, which thus together form a bipolarity through which ex*tends* what Simondon calls the indefinite dyad. This duality, which is also a duplicity, is what Nietzsche posits through the figures of Dionysos and Apollo, and *eris*. And it is what Bergson will reformulate in other terms – by taking thermodynamics and the singularity of life into account in that process which is the "heat death of the universe."

A tendency is a (counter)tendency in relation to another (counter)tendency with which it forms an open dynamic system (which overdetermines a set of transductive relations at a more restricted scale, wherein subsystems form – such as occurs in endosomatic life, where the organism is a dynamic system composed of organs themselves composed of cells). What, in human life, makes the counter-*tendency* a *counter*-tendency? This is the question raised by Bergson, which he describes as the question of a *mystical opening*, which is always in play in any mechanics (ch. 4) – and not just in Newtonian mechanics and its "mechanicism," but in Pindar's sense of *mekhanē* (Stiegler, *The Age of Disruption* 92, 157, 290).

The highly specific situation induced by what is sometimes called "disruption," however, where the question of technodiversity arises as what should come to counteract a state of fact that argues for a single tendency (which is a contradiction in terms, since this would no longer be a tendency but a state – and a state of this kind is well known: it is what we call death, the question here being that of the "mortality of civilizations") – this specificity of disruption lies in the fact that *exosomatic technologies and exomnesic technologies are now integrated* by the *mathematical machines* forming the digital stage of grammatization, thus establishing a stage of exosomatization that is utterly incomparable to those that came before.

## 10 tendencies and entropy

This integration is what characterizes "platforms" (as described by Frank Pasquale, for example) based on the algorithmic, which amounts to the new computational "general equivalent": tending to replace money. But by imposing itself as a state of fact, and no longer as a mere tendency to synchronization *extended out* by its diachronization, the "universality" of this tendency irresistibly generates an anthropic entropy that is fatal for the biosphere-cum-technosphere.

*The challenge is therefore to reintroduce, between exosomatization and exomemorization, the conditions for a variability capable of reconstituting a noodiversity* – a variability of a noodiversity that is obviously also a *technodiversity*, and which alone can cultivate a biodiversity capable of preventing the technosphere from destroying the biosphere. This requires a reconsideration of noesis, its faculties and its functions, from the exosomatic standpoint, and as a noogenesis (see Stiegler, "Postface").

Exosomatic technologies, which we encounter in exosomatization in a way that differs from the endosomatic mode, are a matter of physics and biology, a question of what Simondon called mechanology and what Canguilhem called organology, and belonging to what, after Lotka, we should consider as exosomatic evolution. Exosomatic evolution does *compose* with endosomatic constraints (Canguilhem tries to approach this question in *The Normal and the Pathological*), but it exceeds them, even if it can never completely overturn them: this was, for example, precisely Stephen Hawking's situation, living his life almost but not completely without endosomatic constraints – for if he had somehow completely gone beyond these constraints, he would not have died.

This having been said, and before going any further, let us come to the third point. The technical milieu (in the sense defined in *Milieu et techniques*) is *always in excess* over the interior milieu, and it brings connections to other interior milieus (other "ethnic cells") by entering into *resonance* with their technical milieu – and by traversing their common exterior milieu.

Today, this traversal *saturates* the exterior milieu: there is no longer any external milieu, there is no longer anything except an interior milieu that, consequently, is itself exhausted, devoid of its potential for nourishing heterogeneities, that is, for interior milieus that last up until the moment they are *disintegrated* by connections traversing and linking interior milieus through technical milieus – *the technosphere no longer consisting of anything but a single technical milieu.*

Exomemorization leads to the exorganization of a meeting between three layers:

- a *physiological layer*, as in the case of non-hypomnesic exosomatic organs, articulated with endosomatic organs;
- a *nervous layer*, constituting the cerebral characteristics of the exosomatized, exorganized, hypomnesically constituted being, where all of this presupposes educational models, that is, social rules;
- a *logical layer*, which was for a long time (in the West) a matter of *logos*, of the *skholē*, and of *otium*, but which, with logical machines (themselves based on the microphysical properties of matter), have become a matter of technology.

Between the second and third of these layers, symbolic exorganizations (in the first place, idioms) are woven. The arrangement of these three layers is what characterizes higher complex exorganisms.

The great challenge of these questions will be, particularly in the post-Covid period, to constitute a new theoretical computer science in the service of an economy of negentropy, with a view to reaching the Neganthropocene era, based on a neganthropology, and inscribing, at the *heart* of that theory, the incalculability and irreducible différance of the understanding and reason – playing out across the imagination, where the latter is itself constituted by hypomnesic tertiary retentions, and where Leroi-Gourhan's universal technical tendencies constitute *techno-logical schemas.*

*To be continued.*

## disclosure statement

No potential conflict of interest was reported by the author.

## notes

1 Including (and even usually) by denying it, by default, this denial and this default being symptoms of the extreme urgency of the question.

2 This is what was studied in Stiegler, *Pharmacologie du Front national*, which unfortunately has not been translated into English.

3 This remark is a response to a question from Paolo Vignola.

4 It would be at once indispensable, long, very difficult and no doubt sometimes very painful to delineate this "near" and this "far" in their countless nuances within this very *compromise, whose sense here is not moral but epistemic.* This question, which I outlined in Stiegler, *States of Shock*, and took up in the works that followed, fundamentally involves a reinterpretation of both what Marx posited at the beginning of *The German Ideology* and in the *Grundrisse*, and the question of *Gestell* in Heidegger – a question about which Derrida, as far as I know, always remained strangely silent.

5 In Richard Dawkins's sense.

6 In the sense in which Peter Thiel appropriated this notion put forward by René Girard.

7 Not without problems.

8 The Nasdaq is both an index and the organization of an automated market, whose president from 1990 to 1993 was Bernie Madoff, later sentenced to 150 years in prison, and one of the founders – as early as 1971 – of automated trading quotes.

9 Recurrence is the dynamic principle of what I call the idiotext. On the general notion of recurrence, see Stiegler (*Technics and Time, 3* 144), and, on the idiotext, see Stiegler (*Technics and Time, 2* 64, 243; "Postface" 862–68).

10 A relativity from which it can deviate locally and temporarily, as, for example, a probability constituted by an order or an organization.

11 In the sense of *function integration* described by Simondon: the process of concretization inasmuch as it leads to what he also calls *associated techno-geographical milieus*. On the recent evolution of the latter, which Simondon did not have the opportunity to analyse, see Stiegler, *The Re-enchantment of the World*; *Automatic Society*; *The Age of Disruption*; and *Au-delà de l'Entropocène*.

12 In the sense developed on the basis of the example of Alan Greenspan in the first chapter of Stiegler, *Automatic Society*.

13 And cf. the response by Kevin Kelly, and my commentary in *Automatic Society*.

14 On the relationship between ideology and marketing, see Stiegler, *Pharmacologie du Front national* 11.

15 An earlier version of this text is available in English: see Stiegler, "The New Conflict of the Faculties and Functions."

16 I owe to David Bates, then to David Berry, this connection with Popper.

17 It is one of the great merits of Derrida's *Of Grammatology* that it is attentive to this issue – and to this escape that was then called "logocentrism" and "ethnocentrism." Yuk Hui and I have long hoped to be able to organize a symposium on these questions in China – which could be entitled *Characteristica Universalis, Theoretical Computer Science and Games of Writing*.

18 This is how we refer to the set of digital technologies that constitute a new cognitive function in what we describe as an exosomatic organogenesis of the faculties and functions of reason (see Stiegler, "Le nouveau conflit des facultés et des fonctions").

19 Those discussed by Adorno and Horkheimer, after Benjamin, and without having completely grasped the stakes of what the latter called reproducibility (which is a repro-ducibility: see Stiegler, *Technics and Time, 3*).

20 Spiritual technologies are implemented by Ignace de Loyola and his missions to serve spiritual exercises in response to the Lutheran dissemination of spiritual technologies, firstly through the Bible translated as a book for everyone, *faith or fidelity being redefined in this way as a reading practice*. Benjamin Franklin will redefine this programme through Calvin by guiding it towards the account-keeping books in which ratios will appear, *fidelity being thus redefined as calculation*.

21 And here, it would be necessary to clarify the position put forward in *The Question Concerning Technology in China* on the process of concretization as Simondon describes it with respect to machinic becoming, a point that is all the more crucial since with this concept Simondon responds to Wiener, and to his consideration of the feedback between machines and organisms, that is, with respect to the recursivity that is the subject of Yuk Hui's most recent book, *Recursivity and Contingency*.

22 In this way constituting the course of this exosomatization of exosimples and exocomplexes. On this point, see Giacomo Gilmozzi et al., "Localities, Territories and Urbanities in the Age of Platforms and Faced With the Challenges of the Anthropocene Era," in Stiegler and the Internation Collective, ch. 2, and Stiegler, *Qu'appelle-t-on panser? 3*.

## bibliography

Adorno, Theodor W., and Max Horkheimer. *Dialectic of Enlightenment: Philosophical Fragments*. Trans. Edmund Jephcott. Stanford: Stanford UP, 2002. Print.

Agamben, Giorgio. "The Invention of an Epidemic." *Coronavirus and Philosophers. European Journal of Psychoanalysis* 26 Feb. 2020. Web. 15 Apr. 2020 <https://www.journal-psychoanalysis.eu/coronavirus-and-philosophers/>.

Amsili, Sophie, and Florian Maussion. "L'usage d'Internet dans le monde in cinq chiffres." *Les Echos* 9 Feb. 2019. Web. 15 Apr. 2020 <https://www.lesechos.fr/tech-medias/hightech/lusage-dinternet-dans-le-monde-en-cinq-chiffres-963228>.

Anderson, Chris. "The End of Theory: The Data Deluge Makes the Scientific Method Obsolete." *Wired* 23 June 2008. Web. 15 Apr. 2020 <http://archive.wired.com/science/discoveries/magazine/16-07/pb_theory>.

"L'appel de 1000 scientifiques: 'Face à la crise écologique, la rébellion est necessaire.'" *Le Monde* 20 Feb. 2020. Web.

Barnosky, Anthony D., et al. "Approaching a State Shift in Earth's Biosphere." *Nature* 486 (7 June 2012): 52–58. Print.

Bergson, Henri. *The Two Sources of Morality and Religion*. Trans. R. Ashley Audra and Cloudesley Brereton. Westport: Greenwood, 1974. Print.

Canguilhem, Georges. *The Normal and the Pathological*. Trans. Carolyn R. Fawcett and Robert S. Cohen. New York: Zone, 1991. Print.

Derrida, Jacques. *Of Grammatology*. Trans. Gayatri Chakravorty Spivak. Corrected ed. Baltimore: Johns Hopkins UP, 1998. Print.

Distler, Catherine. "Réseaux globaux et marches financiers: les leçons du krach de 1987." *Quaderni* 12 (1990): 37–47. Web. 15 Apr. 2020 <https://www.persee.fr/doc/quad_0987-1381_1990_num_12_1_1915>.

Floridi, Luciano. "Marketing as Control of Human Interfaces and Its Political Exploitation." *Philosophy & Technology* 32 (2019): 379–88. Print.

Hui, Yuk. *The Question Concerning Technology in China. An Essay on Cosmotechnics*. Falmouth: Urbanomic, 2016. Print.

Hui, Yuk. *Recursivity and Contingency*. London: Rowman, 2019. Print.

Husserl, Edmund. *The Crisis of European Sciences and Transcendental Phenomenology: An Introduction to Phenomenological Philosophy*. Trans. David Carr. Evanston: Northwestern UP, 1970. Print.

Klein, Naomi. *The Shock Doctrine: The Rise of Disaster Capitalism*. New York: Metropolitan, 2007. Print.

Leroi-Gourhan, André. *Milieu et techniques*. Paris: Albin Michel, 1973. Print.

Livet, Pierre. "La notion de récursivité, de la première cybernétique au connexionnisme." *Intellectica* 39 (2004): 125–37. Web. 15 Apr. 2020 <https://www.persee.fr/doc/intel_0769-4113_2004_num_39_2_1214>.

Longo, Giuseppe. "Letter to Alan Turing." *Theory, Culture & Society* 36.6 (2019): 73–94. Print.

Lotka, Alfred J. "The Law of Evolution as a Maximal Principle." *Human Biology* 17.3 (1945): 167–94. Print.

Lovink, Geert. *Sad By Design: On Platform Nihilism*. London: Pluto, 2019. Print.

Morozov, Evgeny. *To Save Everything, Click Here: The Folly of Technological Solutionism*. New York: Public Affairs, 2013. Print.

Nancy, Jean-Luc. "Viral Exception." *Coronavirus and Philosophers. European Journal of Psychoanalysis* 27 Feb. 2020. Web. 15 Apr. 2020 <https://www.journal-psychoanalysis.eu/coronavirus-and-philosophers/>.

Popper, Karl. *Objective Knowledge: An Evolutionary Approach*. Revised ed. Oxford: Clarendon, 1979. Print.

Ripple, William J., et al. "World Scientists' Warning to Humanity: A Second Notice." *BioScience* 67.12 (2017): 1026–28. Web. 15 Apr. 2020 <https://academic.oup.com/bioscience/article/67/12/1026/4605229>.

Simondon, Gilbert. *L'individuation à la lumière des notions des forme et d'information*. Grenoble: Jérôme Millon, 2005. Print.

Simondon, Gilbert. *On the Mode of Existence of Technical Objects*. Trans. Cecile Malaspina and John Rogove. Minneapolis: Univocal, 2017. Print.

Stiegler, Bernard. *The Age of Disruption: Technology and Madness in Computational Capitalism*. Trans. Daniel Ross. Cambridge: Polity, 2019. Print.

Stiegler, Bernard. *Au-delà de l'Entropocène*. Forthcoming.

Stiegler, Bernard. *Automatic Society, Volume 1: The Future of Work*. Trans. Daniel Ross. Cambridge: Polity, 2016. Print.

Stiegler, Bernard, ed. *Digital Studies: Organologie des savoirs et technologies de la connaissance*. Paris: FYP, 2014. Print.

Stiegler, Bernard. *Nanjing Lectures 2016–2019*. Ed. and trans. Daniel Ross. London: Open Humanities, 2020. Print.

Stiegler, Bernard. *The Neganthropocene*. Ed. and trans. Daniel Ross. London: Open Humanities, 2018. Print.

Stiegler, Bernard. "The New Conflict of the Faculties and Functions: Quasi-Causality and Serendipity in the Anthropocene." Trans. Daniel Ross. *Qui Parle* 26.1 (2017): 79–99. Print.

Stiegler, Bernard. *Pharmacologie du Front national*. Paris: Flammarion, 2013. Print.

Stiegler, Bernard. "Postface: Le nouveau conflit des facultés et des fonctions dans l'Anthropocène." *La technique et le temps 1–3*. Paris: Fayard, 2018. Print.

Stiegler, Bernard. *Qu'appelle-t-on panser? 1. L'immense régression*. Paris: Les Liens qui Libèrent, 2018. Print.

Stiegler, Bernard. *Qu'appelle-t-on panser? 2. La leçon de Greta Thunberg*. Paris: Les Liens qui Libèrent, 2020. Print.

Stiegler, Bernard. *Qu'appelle-t-on panser? 3. Déconstruction et destruction*. Forthcoming.

Stiegler, Bernard. *The Re-enchantment of the World: The Value of Spirit Against Industrial Capitalism*. Trans. Trevor Arthur. London: Bloomsbury Academic, 2014. Print.

Stiegler, Bernard. *States of Shock: Stupidity and Knowledge in the Twenty-First Century*. Trans. Daniel Ross. Cambridge: Polity, 2015. Print.

Stiegler, Bernard. *Technics and Time, 2: Disorientation*. Trans. Stephen Barker. Stanford: Stanford UP, 2009. Print.

Stiegler, Bernard. *Technics and Time, 3: Cinematic Time and the Question of Malaise*. Trans. Stephen Barker. Stanford: Stanford UP, 2011. Print.

Stiegler, Bernard, and the Internation Collective (eds). *Bifurcate: "There is No Alternative."* Trans. Daniel Ross. London: Open Humanities. Forthcoming.

## from cosmotechnics to a telos of the particular

jason tuckwell

# TECHNICS AND AGENCY

## *the pluralism and diversity of technē*

In the urgency to articulate a philosophy of technology adequate to our contemporary moment, the Greek concept of *technē* has a contested status. In this context, Heidegger is conspicuous for preserving the concept's originary breadth and the generally positive task he affords to it: for Heidegger, *technē* is part of a bulwark against technology's predations upon nature that have been made possible through *Gestell*. On this account, *technē* functions in accordance with *poiesis* to constitute the poetic mode of dwelling proper to Dasein, yoking it to *phusis* as a regulative constraint. In this sense, *technē* is less a tool than a technique (Heidegger, *Question* 13), governing and guiding our specific manner of being in both thinking and making (34).

In contrast, and following the Aristotelian revisions of the early moderns, there is a more dominant conception of *technē*: in this reading, *technē* is caught up with the instrumentalisation of *phusis* and culminates in a materialist analysis of tools and machines. Today, it is this instrumentalist interpretation of *technē* that is prosthetically attached to human agents as a sort of *homo faber*, a being oriented or primed for tool-bearing.[1] Among other consequences, this programme has inaugurated searching inquiries into the ontological nature of technics. But it is fair to say, that it does not exhaust the complex ways in which *technē* continues to illuminate questions about technology, nor the many senses in which the Greeks deployed the term.

In this essay, I return to the problem of *technē*, where *technē* governs the series of actions and powers to manipulate efficient causation. The historical resources latent in *technē* are well suited to this task, because *technē* conceptualises both actions (*praxis*) and production (*poiesis*) as fundamental to human agency. My intention is to bring some of these traditional, Aristotelian functions of *technē* into contact with contemporary understandings of technics (in Bernard Stiegler and Gilbert Simondon) and especially in light of Hui's cosmotechnical approach, where an adequate thinking about modern technology must proceed through a wholesale reconceptualisation of traditional metaphysical categories to uncover within them fundamental relations to technics. In what follows, a reconsideration of *technē* is used to problematise the relations between technics and agency outside of its anthropic limits.

In order to do so, I return to *technē* in the broadest sense of *skill* or technique and the problem of knowing-how-to-do. For the Greeks, this is to specify differences across various sorts of human excellences and where *technē* is something implied in each of them: whether as the productivity of a craftsperson (in the arts); the practical knowledge of good judgement (in politics); or the theoretical understanding of universals (in philosophy). From this perspective, it is possible to say that the distinction between different sorts of human agents comprise variations of their technical capacity to intervene upon and shape different sorts of processes: to exert a shaping effect on natural materials, on public opinion and on ideas. If this conception seems quaint today, it is because where *technē* has been reduced to *poiesis* it has been detached from its causal reliance on human subjects, to become the "self-productive," indifferent proliferation of modern technology. As Simondon has argued, modern technology requires nothing less than a reconceptualisation of how human agency causally relates to technical systems and a new method for integrating technics and culture.[2]

One way *technē* might offer additional resources for this effort is to specify how technique concerns an agent's capacity to transform efficient causes. For Aristotle, the manipulation of efficient causes via *telos* is closely related to the notion of *function* (*ergon*). As Korsgaard has explained, "function or *ergon* has a wider range of meanings than just 'purpose.' It can be used to mean work or workings or product or characteristic activity" (134–35). As such, function clarifies the sense in which technique concerns particular manipulations of *technē*, in order to act, to work, to produce, and so on.

In what follows, the proximity of *technē* and function constitutes a path to extend how we understand technique to non-human agents. To propose that non-human agents are capable of something like *technē*, would be to specifically query whether the functions of the living already exhibit a sort of *telos* – an emergent capacity for purposive action. In so far as this is possible, it can further be posited that a rudimentary technical dimension must also inhere within non-human, living agents. In order to potentialise this line of inquiry, I will reexamine Aristotle's claim that all living agents exhibit *logistikōs* – the *skill* specific to calculability that is at work in the most primordial, motive forces. However, this will require a rejection of Aristotle's notion that this functional purposiveness is a donation inhering in nature – I will rather argue it is better understood as a synthetic, emergent property. From this I will extract a preliminary premise: *technē* can be used to identify that this sort of functionality is *technical* and belongs to all sorts of living agents. In this respect, it is possible to suppose that agents are coextensive with a sort of artificiality – something synthetic and not given by their primary generative conditions. This concerns a functional arrangement where a modicum of skill or technique is minimally expressed (Tuckwell, "Technē" 1142–43). As such, *technē* can be understood to describe the genesis of an action that is not purely contingent or accidental, but rather transforms these rudimentary efficient causes.

Finally, these reflections should be distinguished from an instrumentalist interpretation of *technē* that is focused upon the series of products generated by a general *poiesis* – any set of external *prostheses*, be they tools, objects or machines. This is not at all to reject instrumental accounts of technics but rather, and in accordance with Heidegger, Simondon, Hui and others, to incorporate this instrumental technical reality with other modes of being. More specifically, my account will focus upon how *technē* might bring to light a mode of technics evident in the purposive functionality of living organisms. I will conclude by returning these reflections on *technē* within the terms that Hui has recommended – namely, to consider how this reconceptualisation of *technē* reproblematises the epistemological background of technics and to submit this to the findings of more contemporary analyses of technology (especially Simondon, Stiegler and Hui). This discussion will focus on two potential sources of technics' concealment with respect to non-human agents: the first follows as a consequence of positing a purposive *telos* that belongs to all living agents. This will raise a number of questions about the ways that agents have been excluded from the "self-productive" evolution of technics, including some commentary on creativity. The second concerns the genesis of technics itself, to ask whether non-human agents might initiate their own technical genesis. This would not only include non-human agents in the category that remains almost exclusively anthropic, but to raise the more speculative proposition that a non-human technical genesis might unfold in a cosmological dimension. This will be to ask whether non-human agents might also satisfy Hui's condition that any well-formulated concept of technics "is always a cosmotechnics" (*Question* 19).

## myth and telos in cosmotechnics

I will begin with a recapitulation of the central problem that is interrogated in Yuk Hui's *The Question Concerning Technology in China*. Hui begins with the orienting claim that an adequate account of the pluralism of technicity remains forthcoming. He brings this to our attention by arguing that a cosmological dimension, animate in the evolution of technology, leads to distinct "local" technicities; this should be contrasted to the consensus that technology is a universal phenomenon. Hui defends this differentiating role of the cosmic, by way of a Kantian antimony between two genetic problems:

> (1) Technics is anthropologically universal, and since it consists in the extension of somatic functions and the externalisation of memory, the differences produced in different cultures can be explained according to the degree to which factual circumstances inflect the technical tendency; (2) Technics is not anthropologically universal; technologies in different cultures are affected by the cosmological understandings of these cultures, and have autonomy only within a certain cosmological setting – technics is always cosmotechnics. (*Question* 19)

In Hui's presentation, this antimony specifically concerns the status of an "anthropological universality" (*Question* 19) within technics. On the one hand, technics as the "extension of somatic functions and the externalisation of memory" (19) is only differentiated by its passage through local conditions, but on the other, technics would be genetically affected by the cosmological understandings of a given culture that would differently orient distinctive evolutionary developments within it.

What disrupts the possibility of an anthropological universality of technics in Hui's account, is that cultural differences operate as shaping functions within the conditions of possibility through which technics comes into being. Myths in this respect are particularly important because they not only constitute a directedness towards the cosmos, but function somewhat teleologically. If this dimension of technics has not been readily apparent it is because a historically dominant myth has been mistaken as a singular and universal explanation for the genesis of technics.

For Hui, *technē* is a crucial point in this analysis, because *technē* is the vector through which a mythic source has been taken as the orthodox universality thought to characterise technics: arguably then, it is not *technē* per se, but the Promethean myth that establishes and underwrites the dominant account of modern technics. What Hui succeeds in showing is that the anthropological universality attributed to technics is grounded within an epistemic field (in Foucault's sense) and an organisation of metaphysical categories that is culturally specific (*Question* 31). It is this series of settings that constitutes a given mythic foundation and that are expressed in the genetic development of technics itself.

It is in this respect that *technē* operates as the myth *de jure* for technics:

> In assuming a universal Prometheanism, one assumes that all cultures arise from *technē*, which is originally Greek. But in China we find another mythology concerning the creation of human beings and the origin of technics, one in which there is no Promethean figure. (Hui, *Question* 14)

The radical potential of Hui's proposal comes into focus here, in so far as it effects a reversal of Stiegler's thesis (48) that a singular genesis of technics underwrites an anthropological universality, rendering culture anterior to technics: we now discover a mechanism whereby cultural differences generate distinct technical systems. Moreover, it is not only that the historical structure of a given culture shapes *a* particular evolution of technology – it is rather that a *progressive series* of technical geneses continually emerges as technics and the grounding episteme of metaphysical and cosmological forces specific to a culture, transform one another.

Although Hui's analysis should be understood to address technics and culture as transindividual phenomena, this move seems to have more particular consequences. This is most obviously because to query the universality of technics implies that human agents are no longer engendered by it in an "anthropologically universal" way (Hui, *Question* 19). As such, where a particular epistemic ground gives shape to technics and culture it also generates differences in individual and disciplinary expressions: "[s]cientific and technical thinking emerges under cosmological conditions that are expressed in the relations between humans and their milieus, which are never static" (18). On this account, whether in scientific and technical thought, or thought and action more generally, "cosmological conditions" will be expressed through each agent's teleological functions which interact with its milieu.

However, the *telos* here in question is not the Aristotelian kind that understands agency to have a universally donated purposiveness (residing in nature or for that matter, culture): what I rather mean to say, is that despite specific cultural settings, the general movement toward the cosmological co-opts each individual's capacity to generate *telos*. To this extent, the scepticism about teleology as a universal donation cannot account for the immanent problem of *telos* as a faculty belonging to the particular. *Telos* is an agent's capacity for *purposive shaping* and is central to the concept of *technē*: it is a functional directedness that enables us to act towards an "end," which gives to actions a purposive shaping force. Our challenge, is to think the particularity of *telos* and its causal efficacy in a way that does not uncritically transgress the terms of modern technics – this means attending first to the distinctions between individuals and transindividual relations, which in Hui's analysis, engender how cosmological conditions give shape to cultural differences.

In order to begin, I will approach this problem by considering the function of innovative or creative ideas in scientific and technical thinking. This will offer a way to illustrate how the *telos* of particular agents might generate an individual expression out of transindividual relations. In order to do so, I will turn to Charles S. Peirce's reflections upon *hypothesis*. According to Peirce, the hypothetical processes of science and technical thought must necessarily proceed through what he calls abductive inference: "All the ideas of science come to it by the way of abduction" (205). Abductive inference is a logical operation that Peirce claims to have extracted from Aristotle's *Posterior Analytics* (205), which in addition to induction and deduction, constitutes the third logical operation endemic to thinking agents: "Abduction is the process of forming an explanatory hypothesis. It is the only logical operation which introduces any new idea" (218).

Peirce has a number of interesting and prescient reflections on abductive inference. Among these is a general theory of creativity, constituted as a limited projective capacity belonging to living agents more generally. On these terms, creativity is not the sort of absolutely free power described by genius or divine inspiration – it is a functional adaptation that Peirce puts on a "gradient" with *instinct* on the condition that it is no longer conceived of as a passive determination of biological hereditary, but to evidence an abductive faculty in non-human agents (227).[3] In Peirce's terms, abduction is a necessary faculty for organisms to achieve anything like a functional efficacy. In this sense, Peirce sees the inventiveness and the possibility of the new expressed in science and technics as a functional modification of a more fundamental property of living agents.

However, Peirce should not be understood to propose a Lamarckian recursion upon an individual's primary genetic conditions. It is rather, a synthetic, non-genetic evolutionary "gradient" comprising something like modifications of *technique*: "technique" would concern the creation and exchange of individual modulations of technical know-how that accounts for non-deterministic behaviour (about how milieus or other agents operate, for example). In other words, knowing-how-to-do involves phenotypical information as a resource for modulating an organism's efficacy that supplements habituated and repetitive behaviours.[4] As such, what Peirce seems to have in mind are modulations of *telos*, that function via a sort of rudimentary power to hypothesise – a finite capacitive field through which innovative solutions are actualised. On this account, the scientific and technical thinking that innovates *abductively* is a modification of a more generic capacity where individual agents behave in functionally responsive and thus creative ways to their environment.

If this foreshadows a crucial issue, it is because the task before us today is to modulate the eschatological directedness encultured into modern technics. This will require a reconceptualisation of both individual and collective actions as well as a more coherent account of how we can purposively intervene in the cosmological settings through which modern technics is developing.

As a preliminary gesture to this end, Peirce's reflections on abduction can be integrated back into the broader terms of *technē* under examination and in a way that might correspond to Hui's broad aims for the project of cosmotechnics. I'll address this in the form of a preliminary hypothesis: can a functional account of *technē*, where its projective powers of *telos* are clarified through Peircean abduction, reveal dimensions of technics that are occulted in the structure of agency, and specifically, in non-human agents?

In order to answer this question, I will leave the analysis of *technē* undertaken in *The Question Concerning Technology in China* that details how the Promethean myth establishes the universal concept of technics. Rather, I will pursue the functionalist aspects of *technē* that Hui has engaged with elsewhere. In a nuanced discussion of the Platonic and Aristotelian complexities of *technē*, Hui's essay "Algorithmic Catastrophe: The Revenge of Contingency," which characterises *technē* as working upon *efficient causality*, states:

> Τέχνη, as Nussbaum has shown, is reason's attempt to overcome contingency. It has nevertheless created a contingency of the second nature [...] The distinction between *tuché* and *automaton* raised by Aristotle has been transformed due to the automatization of reason. The twofold nature of technics (to both overcome and generate contingency) in its relation to contingency can be understood as *pharmakon*, both a remedy and a poison. (139)

I want to consider what is opened in this formulation of *technē*, as an intervention upon efficient causality. In the following account, I will try to show how *technē* not only works to "overcome" contingency while making it elsewhere proliferate, but further *transforms* contingency through a progressive manipulation of efficient causes. In so far as *technē*, in its broadest sense, is a "knowing-how-to-do," it concerns the problem of how an agent's sensitivity, perception and intelligibility gives rise to a resulting action, and whose purpose is to have some impact upon its milieu. This will be to consider whether an abductive or purposive dimension not only gives shape to scientific and technical thought, but implies that a more fundamental technical dimension inheres within all living agents.

## four cause model

If knowing-how-to-do helps us understand the breadth of *technē* it is because it directs us to the problem of how agents deviate, calculate and act upon the primary processes that determine them. This is the context in which efficient causality can be clarified. In establishing these terms, Aristotle is decisive and an account should be made of how he situates *technē* within his expansive version of causality.

For Aristotle, the four causes model describes a complex of interrelations that make categorisation far from simple. Fundamentally, there are four kinds of cause, predicated upon sets of distinctions between the universal and particular, the potential and actual, the immutable and changeable, and into which they form two dynamic groupings or interrelated tensions: the universal pair of the *formal* and *final*, and the particular pair of the *material* and *efficient*. In what are arguably secondary distinctions, the causes are otherwise differentiated by their relative positions and their directionality or orientation. In this sense, the formal and material are paired as the hylomorphic properties that determine the essential nature of beings. In contrast, efficient and final causes comprise a second pairing, responsible for the *purpose* (*telos*) that brings beings into appearance, accounting for both the transcendent unity across all of being (final) and the particular differences manifest within actual beings (efficient).

There is further, and perhaps more fundamentally, an arrangement between the causes at the level of their overall coherence – this pertains to a metaphysical order, that directs all becoming in a general, unidirectional movement. Conventionally, this would seem to be characterised by a passage from universal to universal through the medial particulars, a movement from concrete essences to their perfect realisation (*formal* → *material* → *efficient* → *final*). However, in his reconstruction, Heidegger clarifies that this fundamental Aristotelian causal arrangement has a different metaphysical import. In Heidegger's account, the causal model is predicated upon an interrelation of forces between δύναμις and ἐνέργεια, potential and actualisation. Although the theory of hylomorphism certainly concerns how a universal potentiality begets an actual being via the imprinting of an active form into a passive material, in Heidegger's view the hylomorphic relation is subordinate to a more fundamental dynamism of forces:

> [F]orce is in itself force for doing, for producing; in the orientation toward what is to be produced, there is the reference to what can be produced. The producible is such an out of something. Being out of something is a mode of the from-out-of-which: as the pliant and formable, some kind of material, for example. The essence of force as force for producing also encompasses in a certain way this being "out of something" of producibility. The "out of something" of a thing's producibility is, however, δύναμις τοῦ πάσχειν. This δύναμις (as ἀρχή, as that from out of which something becomes producible) is implicated in the essence of the δύναμις τοῦ ποιεῖν. This is the case not only in general, but rather every particular producing, for example that of an axe, is related to stone, bronze, iron, and the like [...] The result of all this is that force in itself is the *relation* of the ἀρχή τοῦ ποιεῖν to an ἀρχή τοῦ πάσχειν, and vice versa. The essence of force in itself, in terms of its own essence and in relation to this essence, diverges into two forces in an originary way. This of course does not mean that a definite individual force directly at hand consists of two forces, but rather that this force in its essence, that is, being a force as such, is this relation of the ποιεῖν to an πάσχειν *being a force is both as one* – ὡς μία. The relation [*Bezug*] is an *implication* [*Einbezug*]; force is implicating on the basis of its being directed outward and its going beyond. "Force" taken in this way, understood as being-force, is the "ontological" concept of force.[5] (*Aristotle's Metaphysics* 89)

What is of particular interest in Heidegger's analysis, is the emphasis upon an additional element within the hylomorphic relation – here, the hylomorphic relation is internally structured by an interrelated implication of forces. At its essence, force *is* this tension between the directedness towards which two potentials (δύναμις) are differently oriented: force "*for* doing" or "*for* producing." As such, force itself is incomprehensible without the recognition of its essential directedness. It is because of this irreducible directedness that Aristotle speculatively extends this principle to a metaphysical status – a universal *teleological* force responsible for motivating eternal, concrete essences into a series of finite actualisations or becomings. Of course, a profound scepticism about a universal teleological principle is a tenet of modern thought, but it remains worthwhile to recall how Aristotle arrives at this speculative premise:

> It is absurd to suppose that purpose is not present because we do not observe the agent deliberating. Art does not deliberate. If the ship-building art were in the wood, it would produce the same results by nature. If, therefore, purpose is present in art, it is present also in nature. The best illustration is a doctor doctoring himself: nature is like that. It is plain then that nature is a cause, a cause that operates for a purpose. ("Physics" 199b27–32)

To the extent that one can set the universal complications of Aristotelian teleology to one side, it might be permissible in order to retrieve the impetus of empirical observation: in this passage, teleology appears via a speculative attempt to derive a universal agency and purposiveness in nature *from the particular demonstration of technē*: that is, from its power of manipulating *telos*. It is in this sense that *technē* might be understood as a definitive human faculty – a skilfulness to apprehend and deviate the natural processes that determine agents and their milieus. Moreover, in Heidegger's analysis, we might grasp how essential the function of *telos* might be, in so far as this essence is an *implication*, a tensile gap between two directions as the origin of forces for doing and producing. This would appear to locate a rudimentary *telos* in the referentiality of the "for," an originary and essential implication within being that in differing and deferring potentialises the force through which expression and actualisation are made possible. To this extent, *telos* implicates a *technē*-like functionality that enables being to act and produce.

More generally, to set aside the universal complications of Aristotelian *technē* might otherwise be justified in order to render the concept more compatible with contemporary understandings of technics.[6] That is, by reorienting around the immanent functions of *technē* a very different picture of efficient causality comes into view. What is thereby made apparent is that *technē* seems to fundamentally work via a series of progressive exploits of *contingency*. For Aristotle (as it remains for us) "primary" efficient causation is contingency: it is an accident, which spontaneously emerges in *phusis*. But in the reorientation I am suggesting, the breadth of *technē* might better be understood to carry out a set of skilful manipulations or transformations upon the class of efficient causes. As Nussbaum has shown, *technē* has a central preoccupation with altering the fortunes of luck (*tuche*).[7] Aristotle elsewhere describes how agents modulate spontaneity into chance (Aristotle, "Physics" 197a5; Theobald 108). More conventionally, *technē* exploits the contingencies besetting natural forms, developing from them methods of production that can be pedagogically transmitted (*poiesis*). And finally, *technē* undertakes a mastery of involuntary actions to generate choice and rational action (*praxis*) – and it would be possible to go on, in this manner.

In all of these cases, the skilful calculation performed by *technē* leads to a specific and progressive series of changes upon efficient causality itself, in such a way that agents seem to become *ever less determined by* universal causes or primary nature. If there is a pragmatic principle here in effect, it is that whatever is thought to constitute the nature of primary determination (teleological universals, physicalism or materialism, genetic determinism, and so on) *technē* remains the skilful technique, *to operate upon and deviate these processes*. It is in this sense that the *telos* in *technē* might be argued to functionally exceed whatever ontological force is thought to "originally" determine beings in Heidegger or Aristotle.

## the critical role of telos in poiesis and praxis

As such, a further effort must be made to reconceptualise the particular properties of *technē*. This has been first to detach it from Aristotelian universals in order to emphasise how *telos* transforms efficient causes via the essential arrangement of immanent, functional powers. Where *technē* is broadly the capacity of "knowing-how-to-do," this corresponds to a more general technical capability expressed by *logistikōs*, the skill pertaining to *calculative activity* that gives rationality its fundamental properties. For Aristotle, it is *logistikōs* that divides into the fundamental practices of *poiesis* (making) and *praxis* (action). What this will help to clarify is that in the instrumentalist account of *technē*, it is really *poiesis* that is made to be dominant and that contributes to misapprehensions about a universal technics. As such, what must be shown is that *technē* is not essentially the activity of making via imitation or in accordance with natural becoming, but an artificial, technical faculty to apprehend how processes work and to act meaningfully or purposively in response to them.

First, I will try to clarify how the Aristotelian distinction between *poiesis* and *praxis* remains arguably decisive for *technē*. To begin, both are proximate because they concern *telos*; not the *final* cause (the formal, eternal and universal directly apprehended by *epistēmē*), but rather those *efficient causes* concerning variation. That is to say, *poiesis* and *praxis* are both efficient causes, functionally animated by the different kinds of ends they drive at. *Poiesis*, "to make," has an end outside of itself, which is to say that it is a process of production whose purpose is to produce differences into being: this is why its dominant example is material craft, and also why it is especially relevant for the instrumental character of technics, where it presides over the production of technical objects (tools and machines). On the other hand, *praxis*, "action," is an end in itself, so it is a form of deliberation that is proper to moral and ethical concerns: its highest expression is *phronēsis*, a concern for issues of leadership, good living, political action, and so on.

Now, *technē*'s relationship to these activities is more ambiguous than it might have originally appeared, because in its Platonic formulation, "*technē* goes everywhere that *epistasthai* does" (Nussbaum 444), where *epistasthai* is the closest term to our generic designation, "to know," and where this explicitly concerns the sense of "knowing how," in the form of "knowing how *to do*." As such, *technē* is uncontroversially animate within *poiesis* (as in the case of modern technics, where it is frequently characterised as a type or kind of *poiesis*), for it is only through a skilful deviation of efficient causation (a *telos* with an end outside itself) that *poiesis* "makes" – that is, as a mode of imitation of natural production through which it brings difference into being.[8] This is because the *telos* animate within *poiesis*, and which distinguishes it from *praxis*, does not have an end in itself; that is, it is distinguished by the irreducibility of means and ends, so that the *process* of production (the means, i.e., the skilful work of technical agents) should not be conflated with the *product* (the *telos* proper to *poiesis*, where "to make" has its end in the appearance of an artefact).[9] In modern technics, this conflation ends in the instrumentalist reading of *technē*: here the class of technical objects derive their universal character increasingly from abstract structural relations and processes of material production that are progressively alienated from the purposive intentions, the *praxis* or actions of human agents. Among many other consequences, this has been used to justify the alienation of the higher faculties of *phronēsis* – political and ethical deliberation – from the task of modulating the development of modern technology.

In Aristotelian terms, such a conflation breaks the distinction between *poiesis* and *praxis*, a difference maintained by the function *telos* performs. For Aristotelian *poiesis*, its *telos* functions to *actualise* because "to make" is to produce *something*, some resolution or outcome. As for *praxis*, when Aristotle says that it is an end in itself, he means that in this case *telos* remains a sort of *inherence*. This is exemplified in the task to which *praxis* is allotted, namely, the moral and ethical function of *phronēsis*: that is, *eudaimonia*, happiness, the pursuit of a good life (Aristotle, "Nicomachean Ethics" 1140b5–7). That such an activity could have an end outside itself would be a sort of nihilist absurdity (Aristotle, "Metaphysics" 1048b25–27). That is, one would pursue the good life in order to resolve it, as though the goal of life was to *have lived well*, rather than virtue constituting a technique for good living. What Aristotle distinguishes here is quite a different sort of function for *telos*. This is because in *praxis* where *telos* is an end in itself, it is not at all the termination into a product. In this regard, *praxis* shows that *technē* also presides over *a non-instrumental technical process*, where *telos* consistently inheres within and has no object beyond its actions.

What, therein, are the consequences for *technē*? With its profound complications with know-how *technē* is a source of great ambiguity in our understanding of the Greeks because for any rational deliberation a certain skill is required. Even when Aristotle considers *sophia* – the wisdom that marks the attainment of universal knowledge – he begins again by considering the *technite*, the master craftsperson who possesses wisdom if only about the *particular*; it is only at the threshold of the universal that *technē* is explicitly denied access. And if this perhaps underlines a functional superiority latent in Aristotle's formulation, it is because while *technē* remains the rational skill specific to knowing (like in Platonic *epistasthai*), he renders it functionally proximate to the broader and more primitive faculty of *logistikōs*: the calculability that not only precipitates the activities of *poiesis* and *praxis* but more broadly of *ergon*. In this expanded account, *technē* also exemplifies something essential about nature: it is in this sense that *technē* functionally corresponds with how *phusis* is understood to be operate.

## *logistikōs* and the artifice of non-human agency

Despite some recent calls to be done with the category of the human, there are those – Heidegger, Simondon and Yuk Hui among them – who warn against our hastening its demise. One reason to do so might be in the interests of those non-human agents, who have been so decentred and degraded in order to elevate the status of human agency. As such, we might pursue an alternative approach and widen the circle to more broadly acknowledge the functional complexities of non-human agency (Margulis and Sagan 150; Haraway 125). In the present terms, this will require a more careful consideration of how *technē* can be extended within *logistikōs*. Where *logos* envelops the breadth of rational calculation specific to human cognition – that is, the calculative governance of *technē* across all the rational activities (making, deliberating, judging, theorising, and so on) – *logistikōs* is the skill or practice specific *to* calculation: this is a sort of autonomous *technē* belonging to calculation or cognition broadly conceived and is a property that Aristotle reasons must belong to non-human agents:

> Sensitive imagination, as we have said, is found in all animals, deliberative imagination only in those that are calculative: for whether this or that shall be enacted is already a task requiring calculation; and there must be a single standard to measure by, for that is pursued which is greater. ("On the Soul" 434a6–9)

To be clear, Aristotle consistently states that *technē* and *logistikōs* are fundamentally differentiated by the exclusively human faculty of deliberation, necessary for rational and purposive thought. What I am arguing on the contrary, is that one can see in the range of *logistikōs* how a technical functionality could be understood to broadly inhere within living agents. Consider again what Heidegger has demonstrated to manifest from the "ontological force" – that always, already, originary force "for doing/producing." Heidegger clarifies this force is itself constituted by a primary, tensile interrelation, a sort of self-inherence – this tension is between two sorts of directedness towards, one for producing, the other for doing. If there is arguably implied here an additional, synthetic dimension belonging to agency per se, it is in so far as action thus follows from *a potentialisation or enaction of calculative skill* – an existential or primordial capacity for deliberation. This would specify Aristotle's reflection on *logistikōs* where he states that, "for whether this or that shall be enacted *is already a task requiring calculation*" ("On the Soul" 434a6–9; my emphasis). In Heidegger's elaboration, what is opening here as the essence of being is an ontological force whose power is inseparable from *actualising*: the genesis of this power is the implication of two directions, a reference toward a doing and a producing in which a primordial choice is implied as a capacity to differ and defer. In the reconceptualisation I am undertaking, this is the essence of the power of action that defines what is particular to the being of the technical agent.

It is clear that while Heidegger is elaborating here on the essence of Being in-itself, he is not endorsing its application to anything other than human agents. It is rather my suggestion that this analysis can be generalised to all the agents that exhibit *logistikōs* as a way to more schematically conceive of the abductive function that Peirce considers to belong to all living agents. There is implied a *telos*, and something of a deliberative *telos*, in Aristotle's formula that is responsible "for whether *this or that* shall be enacted" ("On the Soul" 434a6–9; my emphasis). The question, as it was for Aristotle, is what engenders this *telos* – this is what continues to require clarification. And since we have become sceptical of universal explanations, it might rather be said that *logistikōs* identifies the *technical nature of calculability* itself, and thus the rudimentary appearance of a functional operator: at its most essential, this operator is evidenced by the skill of choice enabling action. In this sense, even withholding from or following the primary force for doing, would demonstrate a causally deviating *technique*, announcing a living agent of this kind.[10]

It is on these terms that the possibility of a *prosthesis* in non-human agents might be ventured. Where Heidegger's analysis essentialises a tensile gap or in-tension within being, this tension potentialises referentiality itself; either *for* production or *for* doing. In other words, could this referentiality, the directedness of the *for*, function as a prosthetic or proto-prosthetic suture or structure? If we grant the full functional arrangement that Heidegger identifies, there would certainly be potentialised the force *for producing*, even if there was no further development of the skill to manipulate one's *telos* beyond the bounds of the agent in question. To this extent, it is an open question whether this activity should count as *technical skill*: that is, as a technique for directing calculative behaviour. But if we consider the referential potentialisation of "for doing" – for acting as an end in itself – then our problem is to determine to what extent that a *telos* or abductive function is *necessary* for any activity. And finally, while we might essentialise or consider as genetically determined that beings are predisposed to produce and do, we might not so easily determine why they chose *one or the other*, or perhaps in an even more rudimentary way, why they might defer from one or both. Wherever this deficit cannot be accounted for by determined behaviour or contingency alone, we might speculate that a functional technique is thereby demonstrated, even in the calculative processes of the most primordial living beings. This might be used to supplement what Aristotle more clearly had in mind with *logistikōs*: the skilfulness to calculate the organism's sensitive motilities, be they appetitive or perceptual. This skill of directing the "force for doing/producing" within even "simple" living agents, might then identify a potential way to conceive in them a very rudimentary technical faculty.

## do non-human agents have a cosmotechnics?

In the conclusion of *The Question Concerning Technology in China*, Hui offers some recommendations for a future elaboration of cosmotechnics. One of these proposals addresses the problem of non-human agents: "How are we to renew the relation with non-humans after a hundred years of modernisation?" (Hui, *Question* 163). As Hui makes clear throughout the argument, this cannot constitute a return to prior metaphysical categories of nature, but requires their reconceptualisation in light of technics. By way of a conclusion, I will offer some reflections on how the previous discussion might contribute to this task with respect to non-human agents.

Hui has shown us the limitations of *technē* where it functions as the universal Ur-myth for technics. Reduced in this way, *technē* presides over a profound misapprehension where technics is understood to be a universal phenomenon. As a consequence, technics appears detached from the cosmological domain, dominated by its own "self-productive" evolutionary tendency. In the radical terms in which Stiegler formulates it,

> [t]he zootechnological relation of the human to matter is a particular case of the relation of the living to its milieu, the former passing through organized inert matter – the technical object. The singularity of the relation lies in the fact that the inert, although organized, matter *qua* the technical object itself evolves in its organization: it is therefore no longer merely inert matter, but neither is it living matter. *It is organized inorganic matter that transforms itself in time as living matter transforms itself in its interaction with the milieu.* In addition, it becomes the interface through which the human *qua* living matter enters into relation with the milieu. (48)

On this account, culture and more generally the *telos* of human agents, can only modulate technical evolution in a supplementary or peripheral way. This is one series of the problem that a philosophy of technology must come to terms with, and constitutes one formulation of the antimony which structures Hui's investigation.

The second problem is also quite radical, although a certain modesty of style and a commitment to a philosophical ecumenicalism might do something to disguise it.[11] This problem is Hui's own formulation, where the genesis of technics inseparably emerges from an epistemic ground within which metaphysical and cosmological elements are distributed in a culturally specific way. Not only does this refute the notion of a singular technical genesis and the universalisation of technics it establishes, it also disrupts the way this singular genesis determines human subjects. In other words, it also seems to disrupt the universal character of the prosthetic "interface," through which human subjects come to being. By contrasting Prometheanism, with its naturalist metaphysics, with the metaphysical tension between *Qi* and *Dao*, Hui provides a convincing account of how the cosmological coordinates animate in cultures differentiate and shape the genesis of technics. In locating the cosmological terms that make this difference, Hui therefore demonstrates that a non-universal conception of technics implies that "technics is always cosmotechnics" (*Question* 19). This is Hui's alternate formulation of the antimony in question.

However, if this is only the start of a cosmotechnical paradigm it is because it does not detail a series of definitive historical moments where each culture shapes *a* particular evolution of technology – it is a recognition that we are always in the midst of a *progressive series* of technical evolutions, as technics and the grounding episteme of metaphysical and cosmological forces specific to a culture transform one another. It is for this reason that Hui proposes his project in the terms of an antimony where our task today is to continue to reconceptualise our understanding of technics. On this account, false universalism is only one way to atrophy the metaphysical categories: we will need a commitment to an epistemic and critical vigilance if we are to circumvent the effects of modernism through which modern technics is driving us over an abyss. In this sense, a continual epistemological revision attends to the revelation that technics is always cosmotechnics. It is also for this reason we will have to understand anew how individual actions function and exert a shaping effect on transindividual relations – even and perhaps especially where the power of these actions have been demonstrably diminished.

This essay has touched on some particular points where the metaphysical categories problematising non-human agents might be pressed to reveal further implications with technics. I will focus my conclusive remarks on one of these points, which might be posed as a query to Stiegler's formulation of technics in the following terms: is the categorical distinction that defines the "zootechnological relation of the human to matter" (48) away from the "living to its milieu" (48), as singular and particular as Stiegler claims? In simpler terms, is there any evidence that non-human agents participate in what Stiegler calls "zootechnological relations"? (48).

In so far as the zootechnical relation refers to a different material substrate ("organized inorganic matter"), it is worthwhile recalling Derrida's notion of *différance*, which Stiegler paraphrases in the following way:

> [D]ifférance is the history of life in general, in which an articulation is produced (where art, artifice, the article of the name, and the article of death resonate), which is a stage of différance, and which had to be specified. The rupture is the passage from a genetic différance to a nongenetic différance, a "*phusis* differing and deferring."[12] (175)

Thus characterised, it is clearer why for Stiegler technics has a universal character – it is a new order of genesis arising from a nongenetic difference. On this account, the human is constituted by this non-genetic difference giving it a prosthetic, technical nature. However, it is clear from Hui's proposal – whose engagement with Stiegler is extensive and who figures prominently and positively in the terms within which cosmotechnics is elaborated – we nevertheless encounter a rejection of technic's universal character; in terms more sympathetic with Simondon, there is a pluralist conception of technical geneses attributed to their cosmological dimensions. It is also for this reason that there is an open question about how we should understand human difference, in so far as it is no longer singularly specified by a universally organised, inorganic matter.

Nevertheless, it is also clear that Hui is very attentive to the dangers of an uncritical conflation of technics and culture: in other words, the fact that culture has an impact on the evolution of technics does not evidence an individual mechanism through which a constituted being has a direct, recursive effect upon its generative conditions, a point Hui clarifies with respect to Waddington: epigenetic effects are heritable because they work upon the evolutionary landscape through which a specific genus evolves (Hui, *Recursivity* 204). In other words, culture impacts the genesis of technics through a *transindividual* not an individual relationship, and it is through the metaphysical settings of a culture that an individual might come to contribute to a reshaping of technical genesis.[13]

This is to maintain the independence of technical geneses, expressed as a technical tendency operating through the epiphylogenetic process, which maintains its self-productive distinction.

In light of the reflections I have offered in this essay, it is my understanding that an unfolding of the consequences of Hui's non-universalising of technics has certain consequences for whether the zootechnological characterisation of the human is exclusive. To this end, I have undertaken a reconceptualisation of *technē*, which conventionally describes a human-exclusive *prosthesis* that determines the production of artefacts [*poiesis*] to which I have added a functionalist account of technique [*praxis*]. From this, I have drawn on Heidegger and Peirce to incorporate a more general, speculative account of how *technē* enables an exercise of *telos* – a generative series of efficient causal transformations.[14]

Upon this basis, the question has been asked: is technique, as the activity essential to technical agents, something endemic to the living, something that evidences an already operating technical and *artificial* faculty? To begin, the essence of being Heidegger recounts can be argued to locate another site of the Derridean "rupture" between the genetic and the non-genetic. In Heidegger's analysis, where differing and deferring is the implication of directedness between doing and producing, this is held to constitute the very force for actualising beings. In the terms of my reconstruction, this is an account of an originary technics latent in Aristotle's metaphysics that manifests as a being's teleological, "shaping" function, or the emergence of a particular *telos*.

As such, I can now venture a more decisive formulation of my question: can this characterisation of the originary genesis of technical agency be applied to non-human agents? The rigorous terms and the clarity of Stiegler's presentation provide for us some important criteria to pursue this question. First, can the efficient causes of non-human agents be understood to enact something like an organisation of inorganic material? We are mindful again of the restricted terms in which Waddington revives the Lamarckian notion of epigenetic efficient causes, by maintaining distinctions between biological genotype and those of a phenotype determined between the organism and its milieu (Hui, *Recursivity* 136–37). In other words, what we need to show is evidence of an artificial manipulation of materiality that cannot be attributed to genetic determination and contingency, alone.

With these criteria in mind, I turn to the hypothesis pioneered by Lynn Margulis, who demonstrated that primordial bacterial organisms were responsible for the Great Oxygenation Event (37). This event precipitated the third great geological eon, bringing us from the Archean to the Proterozoic and that presided over an entire terraforming of the Earth. Of course, it would be absurd to propose that these non-human bacterial agents had the transformation of the Earth as their teleological horizon. Indeed, it would be as absurd to suppose there was some cosmic synchronisation of individual intentions there at work, than to suggest a purposive intention governs our own technological terraforming of the Earth, some four billion years later. But that is not the question – the question is does this activity evidence a mode of agential intervention that carries out an organisation of the inorganic?

In order to answer this, I will address Stiegler's second criterion regarding epiphylogenesis, where the organisation of the inorganic operates as the mediating "interface" through which a living agent's nature is fundamentally altered (Stiegler 48). In this respect, cyanobacterial organisms not only preside over a terraforming of the planet and an entirely new arrangement of matter: more importantly, their reorganisation of the inorganic functions as an interface that fundamentally alters the nature of the organisms that follow them, precipitating the emergence of multicellular life forms and the genesis of the "arms race" of adaptation and competition that is the study of orthodox evolutionary determinism. Arguably on these criteria, we might consider that cyanobacteria not only preside over an artificial organisation of inorganic matter, but one that precipitates a secondary, *artificial* nature, by functioning as a rupture, a "passage from a genetic différance to a nongenetic différance, a '*phusis* differing and deferring'" (Stiegler 175).

Nevertheless, this "artificial" characterisation rests upon whether the activities of these primordial, non-human agents exhibit a modicum of purposiveness. Again, this is not a question of a universal teleology, where the productive excreta of oxygen by bacteria constitutes anything like a cosmological intention to terraform the Earth. The purposiveness here in question is rather like the teleological directedness that Canguilhem claims is inseparable from the organisation of the organism, through which it strives for "health" (196–97). In other words, is there some sort of nascent cosmological dimension here in play? If *technē* can provide us with a final resource here, it is by invoking one of its most important conceptual resources where *technē* gives explicit expression to the functional correspondence between technics and art. Although there has been little space to elaborate on this point, Peirce's reformulation of instinct is of further interest in so far as it recharacterises deterministic behaviour into a *creative* power of experimentation, distinct from evolutionary processes. On this account, Peirce's technics is very much in the sense of an artisanal experimentation with technique and materials that exemplifies the original remit of *technē*.

As such, while the problem of an evolutionary artifice remains suggestive in Peirce, it finds a far more thorough development in *A Thousand Plateaus*. Deleuze's earlier preoccupation with an aesthetics of intensive signs and signals is developed with Guattari into an evolutionary account of the genesis of "art-technicians" (Voss 100). Here, the expressive functions developed intensively between an individual and its territory or milieu are potentially deterritorialised for the purposes of creative becoming and "there is something of the Cosmos in these more than ample movements" (Deleuze and Guattari 326). In Deleuze and Guattari's account, *expression* performs the functions that abductive inference does for Peirce, in so far as expression works to liberate the behaviours yoked to instinctual biological determinants (332) by generating them into powers of creative experimentation (317). On these terms, the history of art envelops non-human agents (birds, insects, lobsters, and so on), whose series of functional transformations of expressive techniques forms a gradient with the development of the arts qua *technē*: that is to say, an art history no longer indentured to aesthetic objects but elaborated through an evolutionary technics. On this account and to briefly paraphrase, neither art nor technics waits for humans to begin (320).

It is in the terms of a reconceptualised *technē* that these reflections upon the synthetic emergence of art in non-human agents can be used to inform the aims of a philosophy of technics. The effort to uncover the hidden implications of technics in non-human agents might then lead to another potential source of *anamnesis* about technics regarding its long historical association with art. This is not the art encountered via a conventional art history with a grounding in aesthetic philosophy and that returns too quickly to the restricted sense of *poiesis*, but one attending to the originary breadth of *technē*, where art remains a problem about how agents deviate efficient causes in order to create differences. This formulation, concerning how particular agents can give shape to the ever more empowered transindividual forces driving modern technics, has taken on existential urgency in the Anthropocene, for all living agents, human and non-human, alike.

This might be to include in technics even the most rudimentary and primordial of agents, in so far as their agential being is such that it evidences an in-tension – an intensive dimension – a power of differing and deferral as the condition of possibility for any sort of action. It may then be the case that such a fractional generation of possibility is in itself enough, so that all the creative forces for actualisation engender or capacitate a directedness toward becoming-cosmic.

## disclosure statement

No potential conflict of interest was reported by the author.

## notes

I would like to thank Taylor Adkins for his generosity and collegiality in the preparation of this essay.

1 This reduction of the Hellenistic complexity of *technē*, while frequently acknowledged, has not led to a sufficient reconsideration of the term. This is especially where an account of *technē*, affecting manipulations of causal processes, is particularly effaced (Tuckwell, *Creation and the Function*). Arthur Bradley provides an effective summary of this reduction through the Aristotelianism of the early moderns, where *technē* is taken as a foundational resource for a mechanistic concept of nature.

2 This is the major theme of *The Mode of Existence of Technical Objects*, but it is a topic elaborated throughout Simondon's oeuvre. Of particular relevance to the discussion I develop here, are Simondon's recommendations for integrating technics and culture through a common notion of *technique* ("Culture and Technics" 21–22).

3 If abduction is also felt in the genetic conditions of technical genesis, it might be attributable to a logical necessity where evolutionary processes are animated by an inherent and purposive directedness. This is not an innate purposiveness, but a stochastic projective-ness: a finite, probable set of what "might be," which is selective and purposive in order to displace any reliance upon infinite potential or possibility (Peirce 217). This seems to follow from the calculative finitude of living beings. Peirce suggests this is the reason why evolution cannot result from a thermodynamic impetus and pure contingency alone (217).

4 For Peirce, whether in evolution or scientific thinking, habitual and repetitious functions are determined by inductive and deductive logical

processes which appear to be reflexive or compulsive. These seem to constitute the normative constraints that define a particular "nature" of a given disciplinary or living system.

5 This is of course a very different reading of the ontological forces at work in Aristotelian metaphysics than the one Simondon presents in his famous critique of hylomorphism in *L'individuation à la lumière des notions de forme et d'information*. It is beyond the scope of this essay, but a comparison of Heidegger's and Simondon's critiques of Aristotelian metaphysics would be illuminating for both aesthetics and technology.

6 I have elaborated on how Simondon justifies this move in *Creation and the Function of Art* (Tuckwell 89).

7 This is Martha Nussbaum's famous thesis in *The Fragility of Goodness*.

8 This early characterisation of *technē* has been the target of contemporary thinking about technics, especially in Leroi-Gourhan, Simondon and Stiegler, who rather posit that technics possesses "self-productive" powers. While this is a crucial finding, this ontological distinction does not seem to obviate the analysis that finds *poiesis* continues to have a cause outside of itself, which is not solely material in nature but differently resides in the technical agent (*technite*). That racial biases have been built into facial recognition algorithms might serve here as a contemporary example: the self-productive evolutions of computing systems do not surpass but are *complicated* by the causal effects of human *telos*.

9 For example:

> [W]here there are two things of which one is a means and the other an end, they have nothing in common except that the one receives what the other produces. Such, for example, is the relation in which workmen and tools stand to their work; the house and the builder have nothing in common, but the art of the builder is for the sake of the house. (Aristotle, "Politics" 1328a27–35)

10 While this definition of choice – a voluntary action that is exercised whether a *telos* motivates action or not – is Aristotle's ("Nicomachean Ethics" 1111b7–10), I am applying it in a contrary way, according to the expanding reading of *technē* developed across the essay.

11 This point is already amply evident in the methodological underpinning of *The Question Concerning Technology in China*, but a more explicit statement of this commitment can be found in the conclusion of *Recursivity and Contingency*, that returns to the theme of cosmotechnics (230).

12 Stiegler earlier develops this distinction of the human through Derrida's notion of *différance* in *phusis*. Although he is critical of a certain ambiguity in Derrida's formulation, Stiegler more affirmatively draws on this (non)origin of the human, via the concept of *gramme* (playing on grammar and programme) as a break from the biological origin of writing, which constitutes the rupture between genetic and non-genetic programmes (139–41). This is one of the conceptual resources for the "epiphylogenetic" reading of tertiary retentions as constitutive of human difference.

13 Hui quotes a passage from Simondon that clearly expresses this condition:

> By the intermediary of the technical object an interhuman relation is created. That is the model of transindividuality ... The relation to technical objects cannot become adequate individual by individual, except in some very rare and isolated cases; [the relation] can only be instituted under the condition that it succeeds in bringing this collective inter-individual reality into existence, which we call transindividual, because it creates a coupling between the inventive and organizing capacities of multiple subjects. (Simondon qtd in *Recursivity* 176–77)

14 Although this elaboration has been very provisional, this might constitute a method for challenging the naturalist assumptions about instinct, intention and creativity with respect to agency, more generally.

## bibliography

Aristotle. "Metaphysics." *The Complete Works of Aristotle: Vol. 2: Revised Oxford Translation*. Ed. Jonathan Barnes. Princeton: Princeton UP, 1984. Print.

Aristotle. "Nicomachean Ethics." *The Complete Works of Aristotle: Vol. 2: Revised Oxford Translation*. Ed. Jonathan Barnes. Princeton: Princeton UP, 1984. Print.

Aristotle. "On the Soul." *The Complete Works of Aristotle: Vol. 1: Revised Oxford Translation*. Ed. Jonathan Barnes. Princeton: Princeton UP, 1984. Print.

Aristotle. "Physics." *The Complete Works of Aristotle: Vol. 1: Revised Oxford Translation*. Ed. Jonathan Barnes. Princeton: Princeton UP, 1984. Print.

Aristotle. "Politics." *The Complete Works of Aristotle: Vol. 2: Revised Oxford Translation*. Ed. Jonathan Barnes. Princeton: Princeton UP, 1984. Print.

Bradley, Arthur. *Originary Technicity: The Theory of Technology from Marx to Derrida*. London: Palgrave Macmillan, 2011. Print.

Canguilhem, Georges. *The Normal and the Pathological*. Trans. Carolyn R. Fawcett. New York: Zone, 1991. Print.

Deleuze, Gilles, and Félix Guattari. *A Thousand Plateaus*. Trans. Brian Massumi. London: Continuum, 1994. Print.

Haraway, Donna J. *Staying with the Trouble: Making Kin in the Chthulucene*. London: Duke UP, 2016. Print.

Heidegger, Martin. *Aristotle's Metaphysics θ1–3: On the Essence and Actuality of Force (Studies in Continental Thought)*. Trans. Walter Brogan and Peter Warnek. Bloomington: Indiana UP, 1995. Print.

Heidegger, Martin. *The Question Concerning Technology, and Other Essays*. Trans. William Lovitt. New York: Harper, 1977. Print.

Hui, Yuk. "Algorithmic Catastrophe: The Revenge of Contingency." *Parrhesia* 23 (2015): 122–43. Print.

Hui, Yuk. *The Question Concerning Technology in China: An Essay on Cosmotechnics*. Falmouth: Urbanomic, 2016. Print.

Hui, Yuk. *Recursivity and Contingency*. London: Rowman, 2019. Print.

Korsgaard, Christine M. *Self-Constitution: Agency, Identity, and Integrity*. Oxford: Oxford UP, 2009. Print.

Margulis, Lynn. *Symbiotic Planet: A New Look at Evolution*. New York: Basic, 1998. Print.

Margulis, Lynn, and Dorion Sagan. *What Is Life? The Eternal Enigma*. Berkeley: U of California P, 2000. Print.

Nussbaum, Martha C. *The Fragility of Goodness: Luck and Ethics in Greek Tragedy and Philosophy*. Rev. ed. Cambridge: Cambridge UP, 2001. Print.

Peirce, Charles S. *The Essential Peirce: Selected Philosophical Writings Vol. 2 (1893–1913)*. Ed. The Peirce Edition Project et al. Bloomington: Indiana UP, 1992. Print.

Simondon, Gilbert. "Culture and Technics." *Radical Philosophy* 189 (2015): 17–23. Print.

Simondon, Gilbert. *L'individuation à la lumière des notions de forme et d'information*. Grenoble: Éditions Jérôme Millon, 2013. Print.

Simondon, Gilbert. *On the Mode of Existence of Technical Objects*. Trans. Cecile Malaspina and John Rogove. Minneapolis: Univocal, 2017. Print.

Stiegler, Bernard. *Technics and Time, 1: The Fault of Epimetheus*. Trans. Richard Beardsworth and George Collins. Stanford: Stanford UP, 1993. Print.

Theobald, D.W. "Accident and Chance." *Philosophy* 45.172 (1970): 106–13. Print.

Tuckwell, Jason. *Creation and the Function of Art: Technē, Poiesis and the Problem of Aesthetics*. London: Bloomsbury, 2018. Print.

Tuckwell, Jason. "Technē." *Philosophy Today* 63.4 (2020): 1139–53. doi:10.5840/philtoday2020 128316. Print.

Voss, Daniela. "The Problem of Method: Deleuze and Simondon." *Deleuze and Guattari Studies* 14.1 (2020): 87–108. doi:10.3366/dlgs.2020.0392. Print.

# PHILOSOPHY IN THE LIGHT OF AI
## *hegel or leibniz*

sjoerd van tuinen

## ai and philosophical temperament

*All systems are true in what they assert, false in what they deny.* (Walter Benjamin)

Philosophy already has a long history of coming to terms with artificial intelligence (AI). But if the future of the concept is indeed inseparable from artificial languages and ubiquitous computing, then philosophy must also be able to understand and rewrite its own history in this artificial light. Looking back from the vantage point of contemporary AI, how do we interpret and evaluate the various artifacts with which philosophy has sought to cultivate intelligence?

Recently, Yuk Hui and Reza Negarestani have pointed to the affinity between the Hegelian notion of absolute spirit and the functioning of intelligence found in cybernetics and systems theory (Hui), as well as in cognitive science (Negarestani). Hui's approach is historical. He traces a fault line between the organicism of post-Kantian philosophy on the one hand, which marked a clear break with the preceding "mechanization of the world view," and, on the other, the technological organology of post-war cybernetics (the "organized inorganic") and the rise of a planetary noosphere running on algorithms equipped with big data (the "organizing inorganic") (Hui, *Recursivity*). Negarestani's approach is more atemporal. He emphasizes that Hegel's characterization of spirit already came with the awareness of its being the "artefact of a deprivatized semantic space," and that its task has always been to retrospectively recognize its own inorganic conditions of realization (Negarestani 1).

Perhaps we could say that Hui's and Negarestani's approaches to AI relate like the phenomenology of spirit and the science of logic. Regardless of their differences, they share the intuition that the current exponential growth of AI marks a critical threshold in the self-fulfillment of the speculative reason of metaphysics. For Hui, it indicates a rupture between concept and life that goes by the name of "the necessity of contingency" (the equivalent of what Hegel called the cunning of reason); for Negarestani, the history of metaphysics turns out to have been only the prehistory or "prototype" of a posthuman intelligence. More importantly, the approaches share an understanding of intelligence as self-authorizing and self-propelling, that is, as exterior to our entrenched experience of ourselves as human. The "totalizing tendency of far-too-humanist modern technology" for Hui means that technology is no longer an instrument in the human exploration and exploitation of the world, but the cybernetic organization (*Gestell*) in which humans are fated to live. Borrowing the drastic formulation of Negarestani, this inversion of the relation between humans and intelligence raises the question of how critical self-consciousness can "become a part of the veritable history of intelligence"? (3).

I fully agree that intelligence should be seen as having always been artificial in its publicly and linguistically conditioned nature and that, increasingly, what is put into question is the continued relevance of – what is traditionally called – reason to its external, existential conditions. Yet here I would like to propose an alternative to the Hegelian understanding of what it means for spirit or reason to configure itself in the world – an understanding in which there is not one general AI that realizes the unified idea of itself, not even that of technics as such, but one which is capable of divergent becomings, as well as of maintaining and enhancing a stronger sensibility for the multiplicity of historical, material, and technological practices.

Philosophically speaking, this more "well-tempered" understanding of intelligence reaches all the way from contemporary authors such as Isabelle Stengers and Michel Serres to the (pre-Kantian, non-organicist, but perhaps pan-organicist and by all means a-mereological) system of the greatest polymath of the classical age, Leibniz (who, in turn, hoped that through contact with the Chinese, Europeans could rediscover Christianity under modern conditions (*Opera Omnia* 188)). At stake in this alternative lineage is what Hannah Arendt, following Kant, has called "the life" of the mind, that is, the unnatural possibility of a common sense which cannot be realized, but only actualized in the transversal communication and reciprocity of different perspectives. Accordingly, common sense has little to do with the false immediacy of what Negarestani dismisses as the "everyday talk which, lacking objectivity, is in every way arbitrary and dogmatically subjective" (17) and therefore in need of unrelenting critique. It is indeed a multi-agent, externalized and enlarged capacity for thought, maintained and adjusted through regular feedback loops ("habits," in more philosophical parlance), but it is to be approached as a virtual structure with both entropic and negentropic tendencies. As such it is in need of protection against those accelerationists who are ready to "abolish what is given in history" in their Platonic ascent towards "the good as the ultimate vector of cognitive and political emancipation" (504).

This means that I share more sympathy for Hui's call for a multiple cosmotechnics than for the hyper-rationalism of Negarestani. Not unlike Félix Guattari's well-documented fascination with the Japanese "singularity," the combination of technological modernism with pre-industrial cultural traits in Japan (Genosko 122–54), Hui's project is that of seeking "another version of world history," one that is not simply that of the global synchronization with Western modernity, whether in the form of US imperialism or the Cultural Revolution of Mao (with current Sino-futurism as the apocalyptic conclusion of modernity), but rather that of a different integration of technoscience into the continuity of historical cultures, especially non-Western and non-modern ones. It is precisely by "relativizing" what Heidegger once called the question of technology that we can also raise the cosmopolitical question whether Another World, and therefore another framing of the globe, is possible. If the West has lost the cosmos to astrophysics and tends towards an apocalyptic endpoint, perhaps other traditions enable us to retrieve a capacity for thoughtfully participating in a shared cosmos, by which Hui means modes of thinking more integrally and more differentially at the same time. The task of philosophy is "to negate the totalizing tendency in organic thinking" (Lovink,

"Cybernetics") and thereby reopen the future of technology. Only through the affirmation of "technodiversity" does technics become universal. This means that we do not deduce difference from sameness (as in the identity politics of multiculturalism) but induce sameness through the affirmation of differences. "We can relativize a concept in order to universalize, to come to 'the same'" (Lovink, "Philosophy").

In what follows, I will contribute to this project from within the European philosophical tradition by sketching some contours of a portrait of Leibniz following the Deleuzian method of dramatization (Deleuze and Guattari 61, 84). The aim is to displace Heidegger's reading of Leibniz as the first full expression of the cybernetical destiny of Western metaphysics. I want to invite us to consider the possibility that it is precisely from Leibniz that we can learn what it means to desubstantialize tradition (or what we have called "common sense") from nationalism or the culture industry without subordinating it to the perfectionist optimization of science and technology. Is another way of framing digital technology possible, based on a destitution of good sense? At stake in the future of AI is the care for the possibility of a true sense for the common, a sense that is currently everywhere in crisis, whereas critical good sense is always eager to usurp into its own totalitarian image of thought what it simultaneously reduces to a folkloristic obstacle.[1] When we go back to Leibniz, this is not to deny scientific progress, nor to give an exhaustive account of his philosophical system, which is anyway impossible, as the Leibnizian system only exists as the cumulative effect of its various philosophical and mostly non-philosophical models. Rather, the aim is to exhibit, by means of a small number of vital anecdotes that unify thought and affect, the gestures by which Leibniz, as a conceptual persona, moves back and forth between his artefacts for thought and their plane of immanence, which I argue goes by the name of common sense.

## hegel and leibniz

In his *Encyclopedia of the Philosophical Sciences* (1817), Hegel offered the first grand narrative of the self-overcoming of common sense in the form of what I suggest could be described as an enhanced good – mono-directional, self-universalizing – sense that we gradually learn to become the possessors of. Whereas scientific experts provide specialized knowledge, spirit is communal. What renders it absolute is that the community in the full diversity of its historical practices sets the standards of rationality for itself and accepts no exterior authority. At the same time, Hegel finds the most coherent articulation of this internal reflection (or, to paraphrase Hui, the "recursivity" of the negative) at the greatest possible distance to common sense in its immediate givenness. While both science and natural experience get stuck in contradictions between false abstractions ("totalities," as Negarestani calls them), only their infinite mediation by the concept could lead to the concrete and universal self-understanding of the human spirit. The artificial movement of the concept thus constitutes the self-learning system in which common sense is reproduced in its only legitimate form. It is ultimately at the level of the critical reason of philosophy that the judgment over what is and what is not superfluous for the community in its organic totality can be made.

Interestingly, Hegel finds in Leibniz's philosophy the first concept of "contradiction in its complete development," albeit still developed in terms of common-sense logic or "ordinary perception," that is, as "an artificial system" too much rooted in a narrow determination of the understanding (*Lectures on the History* 330, 348; see also Guyer). The contradiction Hegel refers to lies in the antinomous belief that, on the one hand, there is a multiplicity of independent substances, each determined in their complete concept by their own nature, and, on the other, that each substance is related to every other, so that what is true of a single substance is not dependent on this substance alone but in fact on every other substance in the universe. Hegel wants to go beyond what he sees as Leibniz's valiant but vain attempt to reconcile concept and multiplicity, or indeed an analytical rationalism and a synthetical empiricism, by arguing that while contradiction is inherent to the multiple or the finite, the unified concept takes us beyond the immanent limitations of particular things in the more general movement of their dialectical becoming.

And yet the Leibnizian monad is itself already a prefiguration of the Hegelian *Begriff*, perhaps even more. Just as Hegel conceives the self-developing content of spirit as ontological form problem, a monad is the individually determined unity of everything that changes or becomes over time. It is the formal unity of a sensible series determining itself as intelligible "law," a thinking-feeling that perceives and grounds the phenomenal world in its own ideal activity. Both subjective and objective, both for itself and in itself, every monad is a kind of absolute knowledge: both the universal logicizability of the individual and the concept as concrete essence or real efficacy.[2] One could even speak of such a thing as "objective spirit" in Leibniz, to the extent that collectivities of individuals are realized in aggregate phenomena. Composite substances incarnate thought, from organic bodies to the institutions of the state and the church.

The main difference between the post-Cartesianism of Leibniz and the post-Kantianism of Hegel is perhaps that, in the former, absolute mediation or the identity in structure of becoming and thinking is not general but realizes itself only in the individual and in the confederation of individuals. If, for Hegel, this means that Leibniz remains stuck at the level of diversity and particularity, for Leibniz it means that the structure of rationality is never self-present, but only implied. Although both Leibniz and Hegel are pan-logicists, for Leibniz the progressive logic of the mind has the inherently incomplete structure of open-ended resonances. It is regulated through the process of what Gilles Deleuze once called "vicediction" (*Difference and Repetition* 46–80) as opposed to contradiction. Leibniz does not construct his philosophy as the *Bildungsroman* of the subject. As a radically distributed thought, it is composed through elective affinities between irreducible standpoints. Hegel and Leibniz thus stand for two kinds of infinity differentiated in itself, summarized by the following extremes: only the whole is the true vs. only the individual is the true.

## the patron saint of cybernetics

As an example of a cybernetic reason that composes and saves what is given in the singular plural as opposed to what is subsumed under the general, let us therefore turn to that other encyclopedic tradition in philosophy, which I would like to invite you to consider as the potentially more "ecosophical" one. Norbert Wiener regarded Leibniz as "the patron saint of cybernetics" (*Cybernetics* 12). If for Leibniz, monads are mirrors that manage information, for Wiener they are matter-mind/particle wave phenomena. But what is exciting about Leibniz is less that he is the "godfather of the modern algorithm" (Gray), than that, as initiator of the Berlin Academy (where scientists worked together alongside practitioners and to which Leibniz gave the motto *theoria cum praxi*) and firm believer in universal knowledge, he repeatedly identified himself as Pacidius, peacemaker in the tradition of Hermes, the agent of rapprochements who believed in networks to eliminate war and contradiction, not heterogeneity and multiplicity. Where Descartes, looking out of the window, saw nothing but hats and clothes under which might lie automata, and Kant saw integral self-organizing wholes, Leibniz, like cybernetics, sought to reconcile mechanicism and organicism. His encyclopedic project is not the privileged object of philosophy or, for that matter, of any other kind of medium or technology of thought. Excluding all hierarchy between experts and amateurs, rather, it is a referential system for universal communication and circulation, in which each proposition entails every other proposition. To speak with Michel Serres, it is metaphysics as a formal "structure" or "tabular space" (*Le système de Leibniz*), a "site," "protocol," "knot," or a "crossroads" – in short, it is philosophy understood as a "distributor of concepts" (*Hermès I* 1, 2).

If Hegel could still describe the encyclopedia as a circle of circles, Leibniz envisioned a world without a center and without a model. Or rather, a world whose center lies everywhere and has no circumference: the world as infinitely complicated curve, made up of continuities, overlapping invariances and isomorphisms, that appears to be accidental but which must nonetheless be mathematized (*Hermès III* 2, 3). What appears to make self-learning in Leibniz more sophisticated than Hegel – in both senses: more of a sophist and more refined – is that, although both are philosophers of order, he always puts the emphasis on the *ars inveniendi*. In a time of "alternative truths" and argumentative deafness, inventiveness is a sign of intellectual modesty (cf. Wiener's "care and feeding of ideas" (*Invention*)) as well as of challenge and competition. Reality itself is nothing but the feedback effect of the choices between possible worlds that each of its habitants is constantly making. Irreducible to the past, this means that the common is no longer a question of what is but of what can be. The totality of the system only exists as implied by a concept or field of knowledge. Every specialized discipline, laboratory, artistic practice, political or religious institution, and every new technology must be understood as a partial mirror of this implicate order, as a "transversal optic" (Serres, *Le système de Leibniz* 3) that, in its difference mode of producing resemblance, allows us to "see deeper" (Leibniz, *Philosophical Papers* 153), but in such a way that what is anamorphically reflected is not the more universal movement of the Idea but a superdiverse world in endless variation.

Such is the specificity of Leibniz's logic of versions: a model – for example, in mathematics or urban planning or software design – is never an original, let alone an aim in itself, but a further problematization of the common world. It is not some improvised management of uncertainty that, like the derivate market in financial capitalism or the neoliberal university, is always in need of further optimization and self-critique, all the while evacuating from the planned obsolescence or even preemptive destruction of its always much too complex historical context. That is to say, it is not a self-referential demo that mitigates the question of "the best of all possible worlds" by curtailing temporality into short recursive intervals between one version following upon the next in the name of the resilience of any generic system.[3] (NB: things like the university can always be more inclusive, equitable, or just, provided that they never really are.) Rather, every model is a means for returning to life, every technology is a path towards a multiple cosmotechnics.

Systematically closed but combinatorially open, the *Monadology* immediately affirms an all-inclusive, cornucopian world under infinite construction. Comparable to John Cage's permutational work, it is an a-mereological structure based on the dynamic mediation of all by all (see Irrgang). It describes a virtual world of machinic interactivity of which the flywheels are the individuating but not always well-adapted *habitudines*, each spontaneously constituting a perfect world on its own: a "monad" with "no windows and doors," yet simultaneously expressing all the others in a labyrinthine polyphony. In the baroque system of resonances between monads, there is no such thing as a commonplace, neither in the form of the dictates of "smartness" nor in the form of a natural common sense. Instead, history is itself turned into a database for endless "applications." Things and concepts never cease to emerge out of chaos in a grinding, nonlinear fashion.[4] It is not the orderly world of a classical metaphysician but the anarchic world of a mannerist diplomat who has no fear of going beyond common sense but is always confident enough to take care of it.

## leibnizian sophistication

Instead of multiplying common sense in the service of the specialist forms of good sense, as is the habitus of specialist science and post-Kantian philosophy, Leibniz's method was to multiply and decenter good sense in service of a common sense to come. Always ready to meet the demands of endless varieties of good sense, he betted on their convergence in that mostly unlikely structure called universal harmony. Well over two centuries before pragmatism in philosophy, he thus formulated the following constraint for his own philosophical enterprise:

> A metaphysics should be written with accurate definitions and demonstrations, but

> nothing should be demonstrated in it apart from that which does not clash too much with established sentiments. For in that way this metaphysics can be accepted; and once it has been approved then, if people examine it more deeply later, they themselves will draw the necessary consequences. (Leibniz, *Akademieausgabe* 573–74)

This demand clearly runs the risk of a much too clever lack of resistance on behalf of Leibniz, in which case it would be yet another demand for a good sense that, like a categorical imperative of critique, is equally available to all. But Leibniz's demand is first of all a specific constraint that forbids us to pass judgment on what doesn't measure up to our standards and instead enables us to take responsibility for the potentiality of a situation. In addition to demands, there are also obligations, especially those stemming from the consequences drawn by others. "I despise nothing," he used to say in explaining his art of perspectives, "one must always see people from their good side" (Leibniz, *Akademieausgabe* 377). Unlike Kant, for whom existence is a matter of judgment, and unlike Hegel, for whom this judgment became a matter of life and death in the lord–bondsman dialectic, for Leibniz existence was a matter of degrees of perfection. Leibniz does not see an obstacle in the self-maintaining inertia of the masses. Harmony is not taken for granted as a moral starting point that must therefore be destroyed by critical reason, but as the end of a system for which the new is only the means.

Moreover, the inseparability of reason and passion is far from an anti-intellectualist or anti-rationalist attitude. As emphasized by Stengers, it is precisely good sense that risks stupefying thought and blocking the immanent becomings of the common, as it can easily make us indifferent to its efficacy as a ready-made reason to act or judge.[5] It was Kant who reinstalled a dualism between the intelligible and the sensible and blamed Leibniz for intellectualizing appearances as "a mere play of imagination or of understanding" (Kant A271/B327). But for Leibniz, rationality, the cohering of things, does not happen internal to the subject but behind our backs, in the world itself. Which is also to say that without the sophists there would be no philosophy.

Perhaps it was the sophisticated pessimism typical of someone torn between the roles of Renaissance mage and baroque court intellectual that led Leibniz to demand that philosophy should never be opposed to established sentiments. Perhaps it was this sophisticated pessimism typical of someone torn between the role of renaissance mage and baroque court intellectual that led him to demand that philosophy should never break with received opinion. While this demand clearly runs the risk of a much too clever lack of resistance, it is also a specific constraint that forbids us to judge over what doesn't meet up to the model's standard and that enables us to take responsibility for our present becomings and the potentials of which we are made up. Not breaking with established sentiments means saving and redressing them "much as a mathematician 'respects' the constraints that give meaning and interest to his problem" (Stengers, *Invention* 14). In other words, common sense remains a part of the grounds on which we reason. This inseparability of reason and passion is far from an anti-intellectualist or anti-rationalist attitude. On the contrary, it is generic abstractions that risk stupefying thought, as they can easily make us indifferent to their concrete efficacy as readymade reasons to act or judge.

Unlike the dialectical optimism of Hegel, the irenism of Leibniz goes back to humanist theologians such as Erasmus and Melanchthon, but it is equally inspired by the great sixteenth-century discourses on social comportment of Castiglione and Gracián. On the one hand, the brutal reality in Europe – of class struggle, religious wars, and Ottoman invasions – stood in grave contrast to the development of a trade bourgeoisie, the scientific revolution, new techniques of government, and an elitist culture of extraordinariness and excellence. On the other hand, the mannerist world itself emerged out of a desire to civilize the harsh contradictions of the inchoate modern world. Just as the Council of Trent set the terms for the liturgical discipline of priestly etiquette and protocols that effectuate the real presence of Christ in the sacrament, there appeared a whole civic culture of social presence, of effective being in society. War itself acquires an aesthetic destiny, becoming a question of art and technique.[6] If, in warfare, the available options are to attack or defend, at the institutions of the court it is to associate or separate yourself. Here artificiality is not a flaw but an expression of endurance for, and "passibility" towards, incommensurability: a power to be affected by alterity.[7]

Just as contemporary communication technologies arose out of World Wars I and II, moreover, the Thirty Years' War had impelled new ways of articulating the common. With Latin on the retreat as the pan-European language, Leibniz, like Athanasius Kircher before him, hoped to increase our capacity for thought by constructing a new universal language of permutations, variations, and combinations. Rather than referring experience to a model serving as a normative benchmark, his many inventions of the *characteristica universalis*, combinatory algebra, the binary system, and even music are inclusive languages that come flush with a feeling for singularity. That is, instead of replacing vernacular languages with a pure, Adamitic metalanguage of unambiguous meaning, they remain imbricated in the vernacular's infinite and continuous variation, augmenting it by realizing the virtual structure of common sense in an exemplary manner. They do not deny what other languages say but enhance them with new implications (without themselves being implied by these other languages).

For the same reason, Leibniz was interested in the operative languages of hermeticism and magic. Just as the courtly rules of decorum have a projective and conjectural function, signs and images work as transformative gestures, not consolidating ones. Like myths, they have the function of collectively recording a state change, ritualizing into being what they are about. Aimed at practical rather than theoretical problems, their end is in their means. They serve not to claim the truth and expose things to the natural light of human reason – the archetypal gesture of raising consciousness through enlightenment critique – but to decenter them in the scintillating swerve of obscure intuitions. Between signifier and signified, after all, there are all sorts of corporeal exchanges and incorporeal transformations. This explains Leibniz's interest in the *ars magna*, the ultimate general art of Ramon Lull, which he discovered through a group called the Herborn Encyclopaedists. We could say that here the encyclopedic drive becomes a wild schematism that turns the thinking subject into a "spiritual automaton," a cog in the machine of his own collective initiation, processing sign-indications of the possible.

Instead of a simple, even protomythological opposition between noise and information, or between sensibility and thinking, the hermetic philosopher knows that there is a reversibility between the two. There is a layering of the clear and the obscure and of the confused and the distinct. Too much of the same becomes noise, whereas what is really new information falls on deaf ears if context and criteria of pertinence are lacking. All this implies a statistical theory of communication – statistical, because information is distributive, not collective. On the one hand, to articulate is to give a body to the noise that precedes and surrounds meaning. It is to produce mutually implicative and reciprocal bonds between disparate energies in a language that communicates with what otherwise remains mute. On the other hand, noise is also the redundancy of good sense, as is the case in what nowadays reappears as the problem of the confirmation bias in filter bubbles. Only by staying close to the multiplicitous conditions of emergence of sense can we extend the scope of relatively clear knowledge without a totalizing claim. For Leibniz, every text is only a pretext and every communication an initiation. "Those who know me only by my published works, do not know me" (Leibniz, *Opera Omnia* 64–65). By moving in and out of the shadow, we protect what we communicate against the risk of overgeneralization and immediately implicate ourselves within a kind of shared intimacy.[8] The very improbability and self-referentiality of formal languages are warrants against the seemingly natural claims of established, and therefore exclusive, ways of thinking. They disarm the situated as oversaturated by what has already been said and endow it with the possibility of new translations and inner variations.

## an ecology of thought

For Leibniz, the *ars combinatoria* is not a necessary logic but a procedural code of civility at the interface of mind and technology. Different from the scientific object, which is an object of contestation, the technical object is an object of compromise, which is not the same as conformity (Simondon 125).[9] Where the vernacular hinders the imagination, Leibniz's project is the construction of new forms of communication in the informal dimension between established forms of good sense and a common sense-to-be. It situates itself in the noise to which we are exposed and that allows for a cunning revirtualization or "per-version" of the varieties of good sense. How to convince the Holy See to withdraw its damnation of Copernican astronomy? By proposing that there is no absolute space, such that it is equally possible that the earth circles around the sun and vice versa.[10] How to overcome religious strife? By suggesting that the laws of the soul are Protestant, while those of nature are Catholic. Another combination is always possible. Whenever we find ourselves among double binds, our aim should not be to get to the truth as quickly as possible but to invent a new language – perhaps we could say, a queer language in the German sense of *quer* – that maximizes friction in such a way that the various partisans of good sense slow down and effective aggregation in affective learning processes becomes necessary.

A reluctance to speaking in the name of either the love of truth (a reduction of truth to good sense) or the relativity of opinion (a reduction of truth to a naturalized common sense) leads Leibniz to the more tempered "truth of the relative" (Deleuze, *The Fold* 20), or what Stengers calls a "humor of truth (*humour de la verité*, a variation on *amour de la verité*)" (*Cosmopolitics I* 4, 11). There is certainly more humor in tact than in truth. Whereas the ironist never puts himself at stake, the humorist sets aside any prior distinctions between himself and the other. If you get a joke, you're inside the mode of that joke where you start sharing perceptions that lie beyond your own good sense. The joke is an artifact that makes the human capacity for judgment more magnanimous towards other perspectives, including those of nonhuman agents, and, by consequence, to become more intelligent and concrete. In this sense of a constant mutual attuning such that no voice drowns out the other and thereby impoverishes the whole, Leibniz is a radical opportunist. Arguing "from the outcome" (Leibniz, *Textes inédits* 341) rather than from truth, he multiplies principles in such a way that self-definitions, normative ideals, the sense of manifest destiny (aka modernity), or other delusions of grandeur no longer impede the slow process of composition by which we acquire a sense of what might be commonly possible.

This emphasis on how things become, rather than why things are what they are, enables us to read Leibniz as a minor key philosopher in the same sense that Deleuze and Guattari speak of a minor literature or a minor science. Major and minor refer to modes of thinking *and* feeling. Whereas it is easy to discover in Descartes or Hegel major key thinkers who had an optimistic conception of the power of truth and rationality whose heritage is generally claimed by contemporary enlighteners from all over the political and scientific spectrum, Leibniz's so-called optimism is in fact much more pessimistic and moderate. A minor rationality has no model, we should say with Deleuze and Guattari (108), but only a possible world ("a people and an earth to come") that is appropriate to it. Moreover, as he was writing before the rise of secular humanism, Leibniz's idea of cosmopolitanism or civility is not limited to the Kantian version of a global moral community. On the contrary, it concerns a world that is much more inclusive than any overarching law can cover.

Perhaps we should say that philosophy for Leibniz was essentially a form of what Stengers refers to as *cosmopolitesse*. Contrary to a deduction of the world from first principles, but following the calculus of minima and maxima, it is a disciplined attempt "to minimize requirements and maximize obligations" (Stengers, *Cosmopolitics II* 403). Or in Leibniz's own words, to perfect a system that is "the simplest in hypotheses and the richest in appearances" (*Philosophical Papers* 39). However, perhaps what Leibniz teaches us is that perfection does not know an economy, only an ecology. For what is harmony if not correspondence in diversity with no transcending principle or ground? The minimization of requirements (e.g., making do with a few letters or numbers) is achieved through a radical decentering of the theoretical point of view with respect to the diverging interests of the world. It is an operation associated with the empiricist constraint of always beginning from the middle rather than from our habitual selves. The maximization of obligations (never losing sight of the possibility of peace) is achieved through the all-inclusive ambition of rationalism. Everything given in our habitats must have a reason, the principle of sufficient reason says. For common sense to exist, nothing can be left out of the equation, even if this means turning reason into a convention or artificial protocol. After all, reasons, according to the principle of sufficient reason (*principium reddendae rationis*), are not simply given in nature but are founded on prudent judgment and the civic necessity of establishing laws.[11] They must be "rendered" or "returned" to what is already given in the world of appearances – not to render the world "accountable" to the necessities of good sense, as Heidegger thought, but in order to take responsibility for sufficient equity between what we consider rational and the real itself. Reasons co-determine how we fold the world, that is, let ourselves be affected and transformed by it and thus contribute to the activation and generation of further processes of composition.

In the modern world, never mind today's posthuman triumphalism, this artificiality is simultaneously omnipresent and treated badly. And yet if we want to think technology beyond utility and efficiency, our task is to integrate the cosmic order with the moral order as different versions of the same universal process of globalization, that is, the same multiplicity of technicities. What makes Leibniz's cosmopolitesse interesting for us, living in a pancomputationalist world, is that from him we can learn what intelligence comes down to when there are multiple worlds – intelligence understood not in a restricted sense, as privileged forms of knowledge and exclusive immunopolitics, but as generic, even abundant technologies of reasoning and judging about living together that actively contribute to livable habitats. In an artificial milieu in which everything can be counted and all combinations are possible in principle, there can be no particular rationality or privileged state of equilibrium that controls the real, least of all divine intelligence.

It is true that throughout his life Leibniz kept searching for principles of reasoning that could turn thought into a formal symbolic system, a *calculus ratiocinator* in which every controversy could be settled on the basis of calculation. Yet I think we need to take Leibniz's hope that conflicts, grievances, prejudices, and misunderstandings would all give way to consensus with a grain of salt. My aim here has been to make Leibniz figure in a new mirroring relationship, in which his *mathesis universalis* functions not so much as a subsumption of the real under a one-dimensional language that would only communicate truths, but more as a machine for making us think, that is, as a construction that bets on its relevance to the further tuning of a reality made up of opinions and passions. Its basic adage: if you can't beat them, multiply them. Or put yet differently: instead of feeding them to the Algorithm, create a different computational imaginary. At stake for Leibniz as not a critique of what we are but care for how *we* become. If, for Hegel, the restless movement of the concept constitutes an infinite moving away from common sense, for Leibniz, the endless invention of concepts serves their convergence towards common sense. The ultimate question remains which future forms of AI can perform this kind of cosmopolitics.

## conclusion

As we saw at the outset, the question of cosmotechnics is not a question of integrating nature and technology but of multiple ways of integrating culture and technology. According to theology, the cosmos is created out of nothing. For philosophy, by contrast, it is a problem of participation of humans and nonhumans, that is, a body of virtual reciprocity between extremely divergent articulations. Already for the ancient Greeks, the cosmos was a second nature, not the innocence of a first nature but a nature that is morally coded. For Kant (and perhaps Arendt), too, cosmopolitics could only mean the convergence of universal history and the teleology of first nature in a civic nature. But code and institutions are protocols and in this sense the cosmos has always been a problem of technicity. Cosmotechnics thus merely returns to this problem of coexistence by emphasizing the co-constitutive place of technology within it. Following Simondon and Hui, we could say that human, nonhuman, and technical are modes of reticulation that together make up the process of "cosmopoiesis" (Hui, "On Cosmotechnics" 11).

But while Hui's approach is to set up a dialogue between the ontological turn in the anthropology of nature (Descola's ontological pluralism, Viveiros de Castro's multinaturalism) and Simondonian philosophy of technology in order to imagine multiple, versions of this convergence, our aim has been to identify a minor mode within the Western tradition itself. We find in Leibniz a technical activity that has never been "modern," while also having nothing to do with a pre-modern cosmology. Instead, it belongs to a derivative or secondary cosmology that knows no model of perfection, only degrees. Precluding the idea of mastery over nature, Leibnizian cosmopoiesis lies in distributed rather than collective learning processes.[12] This pluralism is also what sets Leibniz apart from Hegel. As inheritor of nonstandard intellectual traditions in Western philosophy, and as inventor of so many artificial conventions and tools for thought, he did not shrink away from the implications of modern technology but embraced them precisely in order to save the diversity of traditions and contribute to sensible commons. His work thus offers a starting point both for thinking AI-diversity and for a cosmotechnical reconsideration of Western metaphysics and technology.

## disclosure statement

No potential conflict of interest was reported by the author.

## notes

1 For a development of this Arendtian argument, see van Tuinen.

2 Leibniz's "law of the series" has the *Sollenstruktur* that exists in fundamental reality and truth only in the lawfulness of its endless modifications (Horn; see also Belaval).

3 Distinguishing between the architectonic orders of the eternal ideas of traditional metaphysical reason and the cybernetical aim for homeostasis of Cold War rationalism on the one hand, and the notion of self-regulating processes of optimization and resilience derived from a systems approach to ecology on the other, Orit Halpern and her colleagues have criticized the logic of demoing as a mode of futurity and optimism espoused in the name of both perpetual growth and impending disaster. At the same time, she acknowledges that critique has itself become part and parcel of the "smartness mandate," according to which the future can be optimized on the condition that collective reflection on historical ends is reduced to the short-term necessity of an ever deeper penetration of computational infrastructure into the environment: "Critique is itself already central to smartness in the sense that perpetual optimization requires perpetual dissatisfaction with the present and the premise that things can always be better" ("The Smartness Mandate").

4 On Leibniz's common-sensism as anarchically opposed to the overarching power of good sense, and as a direct inspiration to Proudhon, see the many references to Leibniz in Daniel Colson's *A Little Lexicon of Anarchism*.

5

> The problem designated by the Leibnizian constraint ties together truth and becoming, and assigns to the statement of what one believes to be true the responsibility not to hinder becoming: not to collide with established sentiments, so as to try to open them to what their established identity led them to refuse, combat, misunderstand. (Stengers, *Invention* 15)

6 Prévost gives the example of Andrea del Castagno's *David* (1450), in which the real triumph is not a representation (the story) but the event of the presentation (the image) itself – that is, the new manner of combining the aggressive gestures of combat into a mutually inclusive individuation (Prévost 87–115, 103).

7 For Jean-François Lyotard, it is precisely the self-referentiality of language, of thinking feeling itself – in the tension between feeling incapable of dealing with the presence of the other that resists our understanding, and the pleasure of finding new words, phrases, and idioms to communicate this experience – that constitutes the possibility of language towards the inhuman (11–18).

8 In the spirit of the undercommons, Erik Bordeleau emphasizes the relevance of the image of secret society in the context of blockchain-based crypto-"networks with consequences." It would be interesting to further explore the system of decentralized ledgers according to monadological logic with both dark and light aspects.

9 In this sense we must defend Leibniz against William James, who presented his own philosophy of compromise (understood as enthusiastic modus vivendi of turning everything into richer consequences and finding prestige in it) as an alternative for Leibniz's artificiality and alleged lack of empiricism (James 15–20).

10 Cf. Hegel's argument that the truth of gravitational force as category of the understanding is the indifference of the moments of its general lawful constitution of reality (*Phenomenology of Spirit* A III, 90 [92]).

11 Serres (*The Natural Contract* 89–92) distinguishes reason (the necessity of truth and innovation) from judgment (the necessity of arbitration and tradition) in a way that is analogous to our distinction between good sense and common sense. At stake is the (Leibnizian) conversion of reason to a new "equation of optimization, symmetry, and justice":

> Starting with the Enlightenment, reason has presided at the court of judgment [...] Now we are witnessing judgment catching up with reason [...] Through a new call to globality, we need to invent a reason that is both rational and steady, one that thinks truthfully while judging prudently. (92–94)

12 In fact, both Descola and Viveiros de Castro are inspired by Leibnizian perspectivism, the former via Émile Durkheim and the latter, like Deleuze, through Tarde (Pelletier).

## bibliography

Belaval, Yvon. *Études Leibniziennes. De Leibniz à Hegel*. Paris: Gallimard, 1976. Print.

Bordeleau, Erik. "Proof-of-Withdrawal: Finance in the Undercommons." *Finance in the Undercommons*. Ed. Pepita Hesselberth and Joost de Bloois. London: Rowman, 2020. Print.

Colson, Daniel. *A Little Lexicon of Anarchism*. Trans. Jesse Cohn. Colchester: Minor Compositions, 2018. Print.

Deleuze, Gilles. *Difference and Repetition*. Trans. Paul Patton. New York: Continuum, 1994. Print.

Deleuze, Gilles. *The Fold: Leibniz and the Baroque*. Trans. Tom Conley. Minneapolis: U of Minnesota P, 1993. Print.

Deleuze, Gilles, and Félix Guattari. *What Is Philosophy?* Trans. Hugh Tomlinson and Graham Burchell. New York: Columbia UP, 1994. Print.

Genosko, Gary. *Félix Guattari. An Aberrant Introduction*. London: Continuum, 2002. Print.

Gray, Jonathan. "'Let us Calculate!' Leibniz, Llull, and the Computational Imagination." *The Public Domain Review* 11 Nov. 2016. Web. <https://publicdomainreview.org/essay/let-us-calculate-leibniz-llull-and-the-computational-imagination>.

Guyer, Paul. "Hegel, Leibniz, and the Contradiction of the Finite." *Philosophy and Phenomenological Research* 40.1 (1979): 75–98. Print.

Halpern, Orit, et al. "The Smartness Mandate: Notes toward a Critique." *Grey Room* 68 (2017): 106–29. Print.

Hegel, Georg Wilhelm Friedrich. *Lectures on the History of Philosophy. 1825–6 Vol. III Medieval and Modern Philosophy*. Oxford: Oxford UP, 2009. Print.

Hegel, Georg Wilhelm Friedrich. *Phenomenology of Spirit*. Trans. Terry Pinkard. Cambridge: Cambridge UP, 2018. Print.

Horn, Joachim Christian. *Monade und Begriff. Der Weg von Leibniz zu Hegel*. Vienna: Oldenbourg, 1965. Print.

Hui, Yuk. "On Cosmotechnics: For a Renewed Relation between Technology and Nature in the Anthropocene." *Techné: Research in Philosophy and Technology* 2.2–3 (2017): 1–23. Print.

Hui, Yuk. *Recursivity and Contingency*. London: Rowman, 2019. Print.

Irrgang, Daniel. "Leibniz – I Ching – Cage. Blind Thinking and Chance Operations." *Dialogos. Ramon Llull's Method of Thought and Artistic Practice*. Ed. Amador Vega, Peter Weibel, and Siegfried Zielinski. Minneapolis: U of Minnesota P, 2019. 270–82. Print.

James, William. *Pragmatism and Other Writings*. London: Penguin, 2000. Print.

Kant, Immanuel. *Critique of Pure Reason*. Trans. Werner S. Pluhar. Indianapolis: Hackett, 1996 [1781]. Print.

Leibniz, Gottfried Wilhelm. *Akademieausgabe*. Berlin: Akademieverlag, 1923. Print.

Leibniz, Gottfried Wilhelm. *Opera Omnia*. Ed. Louis Dutens. Hildesheim: Georg Olms, 1989. Print.

Leibniz, Gottfried Wilhelm. *Philosophical Papers*. Trans. Roger Ariew and Daniel Graber. Indianapolis: Hackett, 1989. Print.

Leibniz, Gottfried Wilhelm. *Textes inédits*. Ed. Gaston Grua. Paris: PUF, 1948. Print.

Lovink, Geert. "Cybernetics for the Twenty-First Century: An Interview with Philosopher Yuk Hui." *E-flux Journal* 102 (Sept. 2019). Web. <https://www.e-flux.com/journal/102/282271/cybernetics-for-the-twenty-first-century-an-interview-with-philosopher-yuk-hui/>.

Lovink, Geert. "Philosophy of Technology in China: Second Interview with Yuk Hui." *Institute of Network Cultures* 4 Sept. 2017. Web. <https://networkcultures.org/geert/2017/09/04/philosophy-of-technology-in-china-second-interview-with-yuk-hui/?pdf=1446>.

Lyotard, Jean-François. *The Inhuman*. Trans. Geoffrey Bennington and Rachel Bowlby. Cambridge: Polity, 1991. Print.

Negarestani, Reza. *Intelligence and Spirit*. Falmouth: Urbanomic, 2018. Print.

Pelletier, Arnaud. "'The Point of View is in the Body': On the Leibnizian Turns of Anthropology." *Leibniz's Legacy and Impact*. Ed. Julia Weckend and Lloyd Strickland. New York: Routledge, 2020. 225–42. Print.

Prévost, Bertrand. *La peinture en actes. Gestes et manières dan l'Italie de la Renaissance*. Paris: Actes Sud, 2007. Print.

Serres, Michel. *Hermès I. La communication*. Paris: Minuit, 1968. Print.

Serres, Michel. *Hermès III. La traduction*. Paris: Minuit, 1974. Print.

Serres, Michel. *Le système de Leibniz et ses modèles mathématiques*. Paris: PUF, 1968. Print.

Serres, Michel. *The Natural Contract*. Trans. Elizabeth MacArthur and William Paulson. Ann Arbor: U of Michigan P, 2008. Print.

Simondon, Gilbert. *On the Mode of Existence of Technical Objects*. Trans. Cecile Malaspina and John Rogove. Minneapolis: Univocal, 2017. Print.

Stengers, Isabelle. *Cosmopolitics I*. Trans. Robert Bononno. Minneapolis: U of Minnesota P, 2010. Print.

Stengers, Isabelle. *Cosmopolitics II*. Trans. Robert Bononno. Minneapolis: U of Minnesota P, 2011. Print.

Stengers, Isabelle. *The Invention of Modern Science*. Trans. Daniel W. Smith. Minneapolis: U of Minnesota P, 2000. Print.

van Tuinen, Sjoerd. "Common Sense: From Critique to Care (Arendt Beyond Arendt)." *To Mind is to Care*. Ed. Joke Brouwer and Sjoerd van

Tuinen. Rotterdam: NAi Publishers/V2_Institute for Unstable Media, 2019. 124–57. Print.

Wiener, Norbert. *Cybernetics: Or Control and Communication in the Animal and the Machine.* Cambridge, MA: MIT P, 1965. Print.

Wiener, Norbert. *Invention. The Care and Feeding of Ideas*. Cambridge, MA: MIT P, 1994. Print.

clive hamilton

# TOWARDS A FIFTH ONTOLOGY FOR THE ANTHROPOCENE

## the ontological cul-de-sac

Those who argue for ontological pluralism aim to weaken the grip of Western "naturalism," Philippe Descola's term for the ontology that begins from the separation of humans from nature so that, by dint of our consciousness or something similarly unique, humans and their diverse "cultures" become radically distinct from nature and its oneness (172–73, 277). This ontology is closely associated with the ideology of the modern era which is centred on the belief that humans can use modern science to unlock the secrets of nature and, with the help of technology, take control of the natural world. Its mood is one of confidence in humankind's power to use our knowledge and technology to create a better future, that is, one of greater ease and prosperity and, implicitly, a more harmonious society to go with it.

Descola contrasts naturalism unfavourably with three other ontologies that have characterised pre-modern societies – animism, totemism and analogism. Yet naturalism's critics rarely specify which of the alternative ontologies should be adopted as a better way to protect the Earth from the ravages of techno-industrial civilisation. This ontological agnosticism avoids privileging one of the alternatives, which would risk accusations of ontological chauvinism. If the problem is cutting carbon emissions, surely at the top of any list, then are we to go to the Amazon forests, to the Australian central desert or to Daoism for the answer? If we choose an animism from the Amazonian Indians (an animism that itself comes in varieties with more or less similarity with animisms from elsewhere in the world), where does that leave the totemism of Australian Aborigines, which in Australia some argue has the answer to the ecological crisis?

Perhaps in the spirit of pluralism we might say that they all have something to contribute to a new worldview that will help us respond to the ecological crisis. If so, which elements of them can contribute and which of the drawbacks of naturalism should they substitute for? Are we entitled to pick and choose from distinct ontological worlds, combining the "good bits" of each to provide a formula like: modern medicine + political liberties + totemic identification with other creatures + worshipping spirits in inanimate objects = nirvana? Or are ontologies organically woven into social and cultural practices from which they cannot be separated so that, as modes of being, they possess the whole person?

And if we select a pre-modern ontology then how exactly can it inform us about responding to climate change? If the objective is to reduce carbon emissions then, apart from the general injunction to treat nature more respectfully, what can pre-modern ontologies tell us about the choice between wind power, nuclear energy, energy efficiency and carbon sinks?

Today's custodians of pre-modern ontologies are typically less confident than well-meaning pluralists that their traditional practices and ways of seeing the world have the answers to techno-industrial destructiveness. When delegations from the Pacific islands arrive at international climate change conferences, their message is not "our ways are the solution to this problem, listen to our wisdom." Their message is: "You, the leaders of industrial countries, know what the solution is, so do it and do it fast because we are going under."

Guilt over colonialism's dispossession and violence casts a long shadow over the West's intellectual life. Ontological pluralism's search for answers in *difference* displays an admirable respect for other cultures and ways of doing things. Guilt and atonement are the proper sentiments when confronted with enormous crimes, but they are a poor guide to how humankind should understand and respond to the Anthropocene.

I am suggesting that an ontology appropriate to the Anthropocene cannot be found by repurposing pre-modern ontologies, or revitalising those that survive in the crevices of naturalism's vast transformations, and then blending them in some yet-to-be-negotiated combination. Apart from the question of which ontology to select from the smorgasbord of non-Western cultural offerings, there are two other awkward questions for ontological pluralists.

First, there is the sheer impracticality of applying pre-modern thinking to techno-industrial economic systems. Zero emission forms of energy such as wind farms rely on a complex mining and minerals processing infrastructure, sophisticated technologies for their manufacture and a high-voltage electricity grid to get the product to consumers. Going off-grid with solar plus batteries is no less technology, manufacturing and mining dependent. There may well be a partial exception in the case of agriculture and land management, where certain pre-modern *attitudes* to the land, and practices like traditional use of fire for landscape management (suitably modified) could and should be drawn on (Steffensen). But there are seven billion people wanting to eat, so broadscale industrial farming, even with reduced reliance on chemicals, will remain industrial farming. Chickens raised on an organic, free-range commercial farm in Vermont have very little in common with their genetically distant cousins scratching around a village in Nepal.

Second, while naturalism as a way of understanding the human–nature relationship underpins or is consistent with techno-industrial exploitation of the natural world, it's also true that techno-industrial society has generated the strongest criticism of that exploitation and the ontology that goes with it. The grounds for those criticisms have at times borrowed from pre-modern ontologies, mostly eschewing their magical or mystical sensibilities (examples might include the work of post-humanists, romantics like Aldo Leopold and contributors to this journal). Other, more politically potent, criticisms of naturalism have not looked to the pre-modern but have responded directly to contemporary facts – that resources are finite and dumping wastes in land, sea and atmosphere damages our health.[1] These critics – everyday environmentalists found in suburbs across the rich world – do not declare themselves against modern science, technology as such, or the subject–object split. Instead, they want their societies to adopt more sustainable practices and use modern science and technology to ensure that a more thoughtful subject gets on better with the exploited object.

Perhaps the argument so far is in the wrong register, because ontologies↓modes of being giving rise to "understandings" of the nature of reality and one's place in it↓are necessarily abstract. However, thinkers like Eduardo Viveiros de Castro, Donna Haraway and

Bruno Latour write with a view to finding the *answer* to the practical problems of the world today. And ontological pluralists follow a long line of anti-Cartesian critics linking the abstractions of Western naturalism (such as subject–object dualism) to the actual practices that are changing the climate and extinguishing species.

This line of causation from ontological foundations to destructive practices is the basis for the view that new practices require new foundations. Several large steps separate modes of being from social goals and practices. Ontologies and their associated ways of living develop in an opaque and dialectical way. One cannot invent a new ontology and then promote it as a way of changing society; believing one can do so is a modern conceit. Taking as a guide the emergence of modern naturalism in Europe that followed pre-Enlightenment analogism – roughly speaking, from alchemy to chemistry – ontological revolutions take place in leaps and bounds over extended periods, a century or more, and are driven by shifts in intellectual worlds in close collaboration with the changing social and political conditions that contain them.

Therefore, I am not looking for an ontology that "provides the answer"; I am searching for an ontology that we can expect to speak to the meaning of the arrival of the Anthropocene. As the ecological crisis deepens over the next decades, we can expect to see a dialectical process of social–political change and a profound shake-up in philosophy. I'm suggesting that the conditions of life will be transformed in a way that renders all existing ontological understandings anachronistic, and we will be groping towards the elements of a fifth ontology beyond the four described by Descola, one rooted in the radically new dispensation brought by the Anthropocene rupture, a dispensation that destabilises all previous understandings of the human, of nature and of the relationship between the two. And I will suggest that by entering a state never before experienced in its 4.5-billion-year history, the Earth is now something without ontological precedent.

The ontological shifts that accompany the most profound events in the history of any civilisation take time. Viewed historically, the physical impacts of the Anthropocene have come crashing in on humans with an extraordinary suddenness. Expert awareness of global warming dates back not much further than the 1980s (the first Intergovernmental Panel on Climate Change report was published in 1988), and awareness among the broader public and elites is much more recent. As a category of social scientific analysis, the Anthropocene is little more than a decade old. Philosophy has, for the most part, been missing in action. (Theology even more so.) We'd expect a fifth ontology to emerge over the next decades as a new generation of thinkers, in response to changing conditions, formulates it, debates it and pushes it forward. It may be radically opposed to naturalism, but it will be extruded from the new conditions of the Anthropocene and not from Holocene worldviews.

## the advent of the anthropocene

Before I elaborate on this claim, it is essential to understand what the Anthropocene is *and what it is not*.[2] It's a concept based solely on Earth System science. Note that by making Earth System science the starting point I am taking an unapologetically Modern (naturalist!) standpoint, rejecting all relativisation of science and asserting an epistemological distinction between science and social science. As I have argued elsewhere, the situation is too serious, the danger is too great, for us to indulge in post-modern intellectual games (Hamilton, "Too Serious"). With climate science for years under ferocious attack, effectively enabling the climate crisis to occur, social scientists have an ethical obligation to refrain from casting doubt on the expert advice of climate scientists.

The idea of the Anthropocene was not possible before the development of the concept of the Earth System, one that began to emerge in the 1970s and reached its full development in the 1990s (Hamilton and Grinevald). The Earth System is the Earth taken as a whole in a constant state of movement driven by interconnected cycles and forces, from the planet's core to the atmosphere and out to the moon, and powered by the flow of energy from the Sun. It is a single, dynamic, integrated system, and *not* a patchwork of ecosystems. It is this system as a whole whose functioning has been disrupted. Disturbance of the atmosphere is also, unavoidably, disturbance of the other spheres – the biosphere, the hydrosphere, the cryosphere and the lithosphere. Those who use the term "Anthropocene" as a shorthand for "human impact on the environment" (which stretches back thousands of years) are not describing (or visualising) the same phenomenon and misuse the term. They aren't describing the same *object* (the Earth System) as conceived by the leading Earth System scientists.

Steffen et al. described the Anthropocene as a "no analogue state," that is, one without parallel in the geological history of the Earth (*Global Change* 74), a singularity whose unique nature upends the rules understood to regulate a typical interglacial period. So to repeat, the idea of the Anthropocene sprang into existence not as another term to describe the extent of human impact across the landscape, the environment or ecosystems, but emerged to capture a *very recent* change in the nature of human impact on the Earth as a total entity, activity of a kind and on a scale sufficient to shift the geological evolution of the planet. This is the meaning of the canonical statement of the new epoch: in the Anthropocene "the human imprint on the global environment has now become so large and active that it rivals some of the great forces of Nature in its impact on the functioning of the Earth system" (Steffen et al., "The Anthropocene" 843). Human activity has become a planet-altering *force of nature*, capable of shifting the Earth's geological evolution. Human activity is now one of the great processes that drive planetary change, like orbital forcing, plate tectonics, volcanism and natural evolution.

To grasp the idea of the Anthropocene, it's necessary to make a conceptual leap from conceiving of environments and ecosystems to picturing the Earth System as a single, functioning, dynamic totality. Many scientists, even some who write about the Anthropocene, remain within a mode of thinking that predates Earth System science. Although the Anthropocene is described as a new epoch to be added to the Geological Time Scale, it is in fact not only a question of stratigraphy. It describes a shift in the state of the Earth System, albeit one that will in future be observable in the rock layers.[3]

As an aside, Bruno Latour and Tim Lenton object to relying on the idea of the "Earth System" because it begins from a conception of "the Globe" and they want to insist that there is no "whole," because a whole is looked at, judged and controlled from the outside, rather than inhabited from within. There is no whole "above" the parts, they insist, only overlapping elements. It's an important caution against the strong human tendency to posit an outside force above that controls the whole, whether it be God, Providence, Evolution or the idea of Gaia itself, which can easily fall into this trap.[4] However, I want to push back against this argument by suggesting that Latour and Lenton have elided the thing with mental projections onto the thing and that they are throwing out the thing with the projections. There is a third approach that avoids both an outside force pulling the strings and a determined focus on the local that loses sight of the Earth System as a totality (without which there is no Anthropocene). We can think of a whole as consisting entirely of interlocking processes, with nothing beyond the processes, so that the Earth is never a fixed object, but an entity comprised entirely of processes, processes that can come together to create many different states of the whole. In the Anthropocene, the Earth consists of many processes, including human activity among the "forces of nature."[5]

## the new ontological status of nature

The advent of the Anthropocene is a *rupture* in human history and in Earth history. Beginning from around 1950, for the first time in the history of the Earth and for the first time in the history of humankind, a conscious willing being shifted the geological evolution of the planet. As a result, the ontological nature of the Earth as an object changed. No longer was its state determined by the *blind forces of nature*; its state would now be influenced by the *decisions* of an intelligent being. For the first time in Earth's geological history, volition plays a role in the state and future evolution of the planet. Humans have brought about an ontological as well as a physical rupture in Earth history.

Technology-enabled humans are a new *force of nature*. Although typically operating in a dispersed way with no central control, technological activity can be guided by conscious decisions on the part of humankind, or at least large parts of it. So, some nations have decided against nuclear power and GMOs (genetically modified organisms); ozone-depleting chemicals were banned by international treaty; and meddling with human genes is outlawed by convention. Although better known for their failure, international climate treaties have slowed the growth of emissions and stimulated more research and investment in alternative forms of energy.

It's true that other living creatures have changed the functioning of the Earth System, notably cyanobacteria, the efflorescence of which 2.4 billion years ago oxygenated the atmosphere with a massive impact on the state of the planet. But cyanobacteria could not have chosen to act otherwise; they had no will. While cyanobacteria transformed the Earth System as a whole unconsciously, humans have done it in the knowledge of what they are doing, even if some deny, ignore or downplay the knowledge.

The fact that the Earth's geological evolution is now influenced by a conscious force is of the most profound philosophical significance. Previous ontological systems were based on one of two characteristics of humans. The first are ontologies that grew out of humans-as-social-only displaced from the natural world, and arising from their rationality, their consciousness, their will, or their relationship with "being" or God. The second arose out of the relationship of humans to the environment that surrounded them, often in the context of larger natural phenomena such as the arc of the sun and the elements. Even so, the relationship was never one with the Earth as an actual totality, an integrated, functioning system. Nor was the relationship one of physical transformation; it was a spiritual or magical relationship.

The distinctive ontological status of the Earth in the Anthropocene implies a radical shift in how we conceive of the human as a being↓as a creature, the only creature, capable of affecting the geological evolution of the planet, and of doing so consciously. This sets up an entirely new relationship between human beings and the Earth, one that was unknown to all previous ontologies. It cannot be equated with "being connected with nature," living at one with the environment or identifying with animals. Indeed, human activity aimed at making us independent of nature's constraints gave us the ecological crisis.

Despite the suggestions of post-humanist writers, the capacity of humans to change the functioning of the Earth System is not proof that the separation of subject and object was a Western illusion and the definitive repudiation of Cartesian dualism. It is the opposite – proof that the subjectivity of humans is so great and so alienated from nature that it now disturbs the planet's operation. Techno-industrial humans have lost none of their subjectivity; arguably, if subjectivity is thought of as the autonomous capacity to affect one's environment, we (collectively) have more.

Anthropocene Earth is increasingly marked by warming, ocean acidification, extinctions and extreme events that are hostile to human life. Earth System scientists struggle to express the new character of the Earth, reaching for metaphors like "the awakened beast." In the midst of a wave of catastrophic bushfires across Australia, the indigenous historian Bruce Pascoe said we should be "calming the bush down."[6] The defiant reaction of the Earth to human interference in its functioning is changing the way we think about it. Much environmental philosophising about the Earth regards it as the *victim* of human exploitation and rapacity.[7] Notions of the Earth as passive and fragile, including the loving Mother Earth and Pope Francis's "sister who cried out to us," no longer fit with the violent, angry behaviour of the Earth in the Anthropocene.

The same goes for Heidegger's critique of nature as "standing reserve," which imagines nature as always passive, waiting, ready-at-hand, to be used by humans as their instrument (322). Heidegger identified the essence of modernity as "a technically mediated will to power [that] has made it its task to shape the world according to its own vision and for its own purpose," in Ruin's words ("The Destiny of Freedom" 294). But now we see the impossible conceit of this modern project. For Heidegger, "the subjectivity of man reaches its completion as a domination or mastery of the earth" (Ruin, "Ge-stell" 189). As it turned out, humankind's subjectivity reaches its apogee in its capacity to shift the Earth from its geological trajectory. But Earth has not been rendered compliant and controllable; it is defiant and irrepressible. The Earth will always have the last say.

The essential belief of the ontological turn in anthropology is that if we could somehow bring about the return of pre-modern, non-naturalist ontologies then humans would adopt a more respectful, less rapacious and more sustainable way of relating to nature. That may or may not be true, if such a return were possible. Yet the "nature" we would be relating to is no longer the nature it was when those ontologies, and modern naturalism itself, emerged. *And it never will be again* because tipping points have been crossed and unstoppable processes have been set in train.[8] Glaciers are not going to refreeze, ocean acidification is not going to be reversed and extinguished species are not going to be brought back. The Earth cannot (at least on the scale of human history) be returned to its Holocene constancy. It will become increasingly wild and cruel.

In short, if the motive for ontological pluralism is to provide a less destructive alternative to a nature-subordinating naturalism, then going to other cultures for the answer is a mistake. Naturalism's successor will need to confront the "untamed beast" that is the Earth in the Anthropocene. A belief in humankind's power to exercise control over nature may have had some justification in the stable and clement conditions of the Holocene, but it becomes absurd on the increasingly unfamiliar Earth of the Anthropocene (Hamilton, *Earthmasters*, ch. 3). However, this does not mean that humankind has had its agency neutralised; the advent of the Anthropocene in fact proves the reverse. Humankind is enormously powerful, so powerful that it can shift the geological course of the planet as a whole. As a statement about humankind's moral superiority, anthropocentrism may well have been catastrophic; but as a statement about the unique power humans hold among all creatures it is undeniable.

We therefore agree with Yuk Hui's conclusion that "the concept of 'indigenous ontology' itself has to be questioned [...] because it is situated in a new epoch and transformed to such an extent that there is hardly any way to go back to it and restore it" (8). He argues that the answer lies in "a cosmotechnical thinking from the standpoint of these ontologies without falling prey to an ethnocentrism," because all societies, European and non-European, have been transformed by modern technology. But isn't focusing on modern technology missing the essential point of the Anthropocene, that is, the rupture in the Earth's functioning? If the Anthropocene has brought a shift in the ontological state of the Earth, then Hui's cosmotechnics also may miss the essence of the new times. The play between technology and ontologies, subtle as it is, does not come to terms with the profound shift represented by the advent of the new epoch. Yes, technology, colonialism, inequality and the structure of capitalism all came together to produce the Anthropocene, but they were all acting on the world for two or more centuries before the new epoch arrived. It's the rupture in the Earth System and the rethinking of humans as Anthropocene creatures that is now the essential question.

Hui provides a valuable corrective to the belief that humans engage with a "pure and innocent nature" prior to and outside of technology, but his aim to develop a cosmotechnical thinking for the Anthropocene from a lineage of non-Western thought lodges thinking about the Anthropocene in pre-rupture history (20 n. 4). Hui does not reify technology in the way Peter Haff does in his concept of the "technosphere," in which humans "are a product of a system that operates beyond our control and that imposes its own requirements on human behavior" (127), but he does see technical activities as the vector that unifies the cosmic order and the moral order. We do need an ethics of the Anthropocene. While Hui's intermingling of the moral with the cosmic is superior to Haff's abolition of the moral through his reversal of the human–technology relation – so that humans are helpless participants in the autonomous technosphere – doesn't the Anthropocene's remaking of the human as a planetary force invalidate previous conceptions of the moral order, both those that seek to bury humankind's uniqueness in a bioethics of nature and those that make man master of his environment? As I have remarked elsewhere: "An ethical framework that can tell us whether it is wrong to overstate our travel expenses cannot tell us whether it is wrong to change the Earth's geological history" (Hamilton, *Defiant Earth* 148).

The world, the Earth, is so fundamentally transformed in the Anthropocene that it renders all previous ontologies (as modes of human relations with the natural world) obsolete, and it does so for the most basic reason. All previous ontologies assumed their character in the distinctive conditions of the Holocene,[9] conditions that have been consigned to the Earth's record book. It's for this reason that we need (or more accurately, will have thrust upon us) a new way of understanding what it is to live on the Earth beyond the four ontologies described by Descola, one that emerges from a sense of the distinctiveness of the Earth in the new epoch and the emerging relationship of humans to it, one of intense engagement but with no possibility of mastery. At a time when everything is changing because the Earth has entered a "no analogue state," searching for answers in Holocene conditions and pre-modern ontologies anchors thinking more deeply in the past.

## a fifth ontology for a new earth

A starting point for a new ontology is the reconfiguration of time that the Anthropocene calls for. In the first instance, we know the Anthropocene is beginning to express itself in freakish and dangerous weather events, events that will become more common and more dangerous in the decades to come. Suddenly, we are thinking about the future in an entirely different way. Instead of being guided by well-defined expectations over the next few years followed by vague and generally benign visions disappearing into the unknown, now the future extends over decades or a century defined by imaginable conditions of climatic stress. A common mental benchmark is 2100 because that is the chosen endpoint for most climate projections. We can now form a picture of life for our children or grandchildren at the end of the century (the expected lifespan of those born today in a Western nation extends beyond 2100). It's no longer a misty but hopeful one of unknown technological improvement and rising prosperity. So the arrival of the new epoch causes us to stretch our horizon of the future in a very personal way. The distressed words of a twelve-year-old girl at a school strike protest caught the mood. "What's going to happen to humankind?" she wept. "What's going to happen to the whole world, if no one does anything?"[10] Anguish about what this century holds has given rise to the disturbing question now frequently asked by young adults: "Should I bring a child into the world? Is the Anthropocene creating a generation that no longer accepts the role of reproducing the species?"

If the arrival of the new epoch forces everyday personal thinking to consider the conditions of life in fifty or a hundred years' time, it has also brought a deeper shift in the substrate of consciousness, the shift required as a result of the convergence of human history with geological history, after human activity shifted the Earth System onto a different trajectory (Chakrabarty). The Anthropocene forces us to consider the human beyond the accustomed scale of human history (decades and centuries) as a force embedded in geological history (hundreds of thousands or millions of years). The carbon we emit today will still be affecting the Earth System in 100,000 years' time and beyond. In other words, humans for the first time enter into deep time, planetary time, in which we find our position within the unfolding evolution of Earth System processes. No previous ontology has had to build itself around this notion of time, that is, around the human as a geological force as well as a historical one. It is not equivalent to the "timelessness" of Indigenous cosmologies or the cycles of eternal return; instead, it unfolds in a single direction marked by a human–nature dialectic of change.

We are accustomed to thinking of an ontology as the product of deep philosophical thought; but, at its base, an ontology is the philosophical articulation of an emotional orientation towards the world, a disposition or attunement that is not a passing affective state, but is a sense that possesses the whole person and makes possible a certain range of emotions.[11] The disposition of modernity is the feeling of distance from the natural world with confidence and hopefulness in human technological capability, perhaps undergirded by a mood of anxiety due to a loss of meaning or rootedness. The mood of totemism might be the deep sense of assurance of living in a state of cosmic order.

In themselves, ontological systems are puny things. Only when a philosophical school articulates an ontology that expresses the mood of the age does it become the ontological expression of the age, that is, the ontological expression of a lived social reality.

So the fifth ontology will be the expression of the emotional orientation, or attunement, of the Anthropocene, one that responds to and is true to the actual circumstances of the new epoch. What is the emotional orientation of the Anthropocene? In order to uncover it there is little value asking those who have spent their lives immersed in the previous epoch, the products of the modern or the post-modern. But we don't need to look far to get a sense of the disposition of the Anthropocene; it is the emotional orientation possessing the millions of school students who have gone on strike demanding radical action in response to a warming world.

Greta Thunberg found herself on the world's stage because she articulates, simply and with passionate conviction, the feelings of millions of teenagers around the world. She expresses their bitterness – "you all come to me for hope? How dare you!" – betrayal – "You have stolen my dreams and my childhood" – and an inescapable fear, fear of the monster that threatens to devour them – "We are facing a disaster of unspoken sufferings for enormous amounts of people [...] I want you to panic. I want you to feel the fear I feel every day."[12] The rage boils over: "You say you love your children above all else, and yet you are stealing their future in front of their very eyes."

Greta rejects the empty talk of hope – "I'm not going to talk about hope" – yet adds "we do need hope." It's not the hope of marketers and politicians, because that kind of hope has for years been used to obscure the truth and justify inertia. Greta's kind of "hope" is the outward expression of an inner resolve to act, no matter how bleak the circumstances and impossible the goal. "The situation is hopeless, we must take the next step," in the words of Pablo Casals.

Beneath the feelings of anger, exasperation and betrayal, the essential disposition of the students is one of dread. They dread the vast transformation that is taking place in the Earth System and what it means for their lives and life on the planet.

A fifth ontology will provide a structure of ideas to explain and give meaning to the mood of the Anthropocene epoch. It begins from acceptance that catastrophic human failure has shifted the Earth onto a new trajectory, one on which the planet is unstable, uncontrollable and much less sympathetic to life. For some, the new Earth creates a new sense of the sublime, of humans shrinking in awe at the power of nature unleashed, a reverential fear of its danger, its rage and, through it all, its enduring beauty, that is, sublime dread.

What of humans in the ontology of the Anthropocene? An answer begins with the collapse of the great dream of the modern age, of humans using their unique capacities to take control of their environments and create their future. The mood is no longer one of power and confidence, of the belief that we can unlock nature's secrets in order to exploit it. The loss of faith in the systems built by modern humankind is a repudiation of techno-optimism because the extreme danger of techno-industrialism has brought us here. We are inescapably enmeshed in the Earth System so that our power is always countered by a more powerful Earth.[13] But nor is it a rejection of technology; for a tech-drenched generation, technology must be at the centre of any attempts to manage the harm we have done. Our task is not to deny our extraordinary power and attempt to abandon it but to work out how to use our power responsibly, that is, how to deploy technology and management practices to reach a reconciliation, to *calm the Earth*.

In sum, a fifth ontology will be built from the two fundamental facts of Anthropocene science: humans are so powerful that we can change the geological evolution of the Earth itself; but we can never tame the Earth and its defiance leaves us facing a dreadful future. Rather than denying human agency by emphasising human integration into the natural world, a fifth ontology accepts that human agency has been supercharged and taken on a geological temporality, even if it has rampaged across the land with self-destructive abandon. Only an ontology built on the crushing responsibility that goes with the power of our technology can save us.

## disclosure statement

No potential conflict of interest was reported by the author.

## notes

1 This is undoubtedly a huge and complex subject worthy of a book in itself, but I would argue that criticisms of exploitation of nature that are rooted in modernity and even in naturalism include those of Rachel Carson, Paul Ehrlich, E.F. Schumacher, the Club of Rome, Herman Daly, Green parties and most mainstream environmental organisations.

2 For an extended argument, see Hamilton, *Defiant Earth*, ch. 1.

3 A point emphasised by one of the leading Earth System scientists examining the Anthropocene, Jan Zalasiewicz, who writes: "The Anthropocene is not about being able to detect human influence in stratigraphy, but reflects a change in the Earth system" (quoted in Hamilton, "Too Serious"). Work on reconciling geology with the Earth System science is just beginning (see Steffen et al., "Stratigraphic").

4 Latour and Lenton also argue that using the idea of emergent properties – properties belonging to the whole but not to any of its parts and which "emerge" when the system reaches a certain state – is a "cop out" because it tries to join the parts up to make a whole.

5 It could be argued that there is no such thing as a force of nature. Volcanism, for example, is a process that occurs in nature whereby molten rock is ejected from the mantle to the surface, with each eruption depending on a variety of specific circumstances. So while there are many volcanic eruptions, one cannot point to a thing called volcanism. We use the term volcanism to describe the collective phenomenon. Human activity is a force of nature in the same way.

6 <radio.abc.net.au/programitem/pgK6D24BR7> minute 5:55.

7 Pope Francis's encyclical *Laudato Si'i* follows this tradition. See Hamilton, *Defiant Earth* 48–49.

8 See, for example, Lenton et al.

9 This will do as an approximation, although it's more complicated. For example, the totemism of Australian Indigenous people emerged in the Pleistocene, that is, over the 50,000 or so years after they arrived in Australia but before the arrival of the Holocene 10,000–12,000 years ago, and therefore took shape in very different climatic conditions (and with different biota and landscape). As far as I know, no one has studied how that totemism was modified as the climate stabilised in the Holocene, which must have brought about far-reaching changes in how Aboriginal people lived, along with evolution in the forms of totemism that governed their understanding of the cosmos and their role in it.

10 <www.abc.net.au/news/2019-05-03/students-strike-around-the-country-for-climate-action/11077022>.

11 This is close to Heidegger's notion of attunement. For a very helpful discussion, see Cowles 53–55.

12 <www.npr.org/2019/09/23/763452863/transcript-greta-thunbergs-speech-at-the-u-n-climate-action-summit>; <www.theguardian.com/environment/2019/jan/25/our-house-is-on-fire-greta-thunberg16-urges-leaders-to-act-on-climate>; <www.democracynow.org/2018/12/13/you_are_stealing_our_future_greta>. She adds: "You are failing us. But the young people are starting to understand your *betrayal.* The eyes of all future generations are upon you. And if you choose to fail us I say *we will never forgive you.*"

13 A fifth ontology will not deprive humans of their extraordinary agency by blurring it into the "agency" of other living things as a way of dethroning anthropocentrism (Donna Haraway). Nor will it abolish our agency by arguing that humans only imagine they have agency but in fact behave according to the dictates of the systems in which they operate (as in Peter Haff's technosphere).

## bibliography


Chakrabarty, Dipesh. "The Climate of History: Four Theses." *Critical Inquiry* 35.2 (2009): 197–222. Print.

Cowles, Darshan. *Thrownness, Attunement, Attention: A Heideggerian Account of Responsibility.* PhD Diss. U of Essex, 2017. Print.

Descola, Philippe. *Beyond Nature and Culture.* Chicago: U of Chicago P, 2013. Print.

Haff, Peter. "Humans and Technology in the Anthropocene: Six Rules." *The Anthropocene Review* 1.2 (2014): 126–36. Print.

Hamilton, Clive. "The Anthropocene: Too Serious for Post-Modern Games." *Clive Hamilton* 19 Aug. 2014. Web. 1 Dec. 2019 <clivehamilton.com/the-anthropocene-too-serious-for-post-modern-games>.

Hamilton, Clive. *Defiant Earth: The Fate of Humans in the Anthropocene.* Oxford: Polity, 2017. Print.

Hamilton, Clive. *Earthmasters: The Dawn of the Age of Climate Engineering.* New Haven: Yale UP, 2013. Print.

Hamilton, Clive, and Jacques Grinevald. "Was the Anthropocene Anticipated?" *The Anthropocene Review* 2.1 (Apr. 2015): 59–72. Print.

Heidegger, Martin. "The Question Concerning Technology." *Martin Heidegger: Basic Writings.* Ed. David Farrell Krell. New York: Harper Perennial, 2008. 311–41. Print.

Hui, Yuk. "On Cosmotechnics: For a Renewed Relation between Technology and Nature in the Anthropocene." *Techne: Research in Philosophy and Technology* 21.2–3 (2017): 1–23. Print.

Latour, Bruno, and Timothy Lenton. "Extending the Domain of Freedom, or Why Gaia Is So Hard to Understand." *Critical Inquiry* 45.3 (2019): 659–80. Print.

Lenton, Timothy, et al. "Climate Tipping Points – Too Risky to Bet Against." *Nature* 95 (28 Nov. 2019): 592–95. Print.

Ruin, Hans. "The Destiny of Freedom: In Heidegger." *Continental Philosophy Review* 41 (2008): 277–99. Print.

Ruin, Hans. "Ge-stell: Enframing as the Essence of Technology." *Martin Heidegger: Key Concepts.* Ed. Bret Davis. London: Routledge, 2014. 183–94. Print.

Steffen, Will, et al. "The Anthropocene: Conceptual and Historical Perspectives." *Philosophical Transactions of the Royal Society A* 369 (2011): 842–67. Print.

Steffen, Will, et al. *Global Change and the Earth System: A Planet Under Pressure.* Berlin: Springer, 2004. Print.

Steffen, Will, et al. "Stratigraphic and Earth System Approaches to Defining the Anthropocene." *Earth's Future* 4.8 (2016): 324–45. Print.

Steffensen, Victor. *Fire Country: How Indigenous Fire Management Could Help Save Australia.* Richmond, Victoria: Hardie Grant, 2020. Print.

frédéric neyrat

*translated by daniel ross*

# THE BLACK ANGEL OF HISTORY

## *afrofuturism's cosmic techniques*

[...] and why not the sun. (Frantz Fanon, *The Wretched of the Earth*)

### introduction: the cosmic techniques of the alienocene

> You can't go into outer space unless you count down to zero. Everything started from zero, you see. You count down to zero, then you can go into outer space. They do it, NASA does it every day, they count down first. (Youngquist 215)

These words by the musician and poet Sun Ra might simply seem absurd: a countdown cannot reasonably be considered a condition of possibility of launching a rocket, since there is no causal link between the first event and the second – as usual, Sun Ra is playing with us. But the jests of this jazz musician, who would come to be considered an ancestor of Afrofuturism, should still be taken very seriously. That there is no causal link between the launching of a rocket and a countdown, or in other words that disjunction lies at the heart of what connects a cutting-edge technology, aerospace engineering, to the span of time preceding a rocket launch, should grab our attention: what is in question is indeed the relationship between technology and time. And this relationship is expressed by Sun Ra in the form of an inversion: before being able to propel a rocket towards infinity, technological power requires a preliminary operation, a descent to zero, a kind of originary negation reducing time to nothing and space to ground altitude.

This non-technological prerequisite – this originary negation and the dialectic that it engages between elevation and descent, universe and planet Earth – circumscribes the blind spot of today's dominant technological apparatus. Indeed, the currently fashionable term for the relationship between technology and terrestriality is "Anthropocene," which posits humankind as having the geo-morphological power to fundamentally transform the Earth. Understood in this way, the Anthropocene should be seen as the symptom of a foreclosure of the *cosmos*, a term that I understand not as world (*Welt*, *monde*), or as a set of persons (human and non-human), but first of all as an expanding universe composed of planets, galaxies and black holes.[1] This foreclosure of the *cosmos* is the effect of a threefold reduction at work in the

Anthropocenic conception: (1) geocentric reduction, since by definition only the Earth, as an object-system, receives geo-morphological attention; (2) anthropocentric reduction, since humanity is ultimately posited as the subject solely responsible for the Anthropocene, in some way rendering opaque the geo-technological and ontological apparatus in which it is caught; (3) leukocentric reduction, since the dominant geo-technology is a white technology (*leukos*, in Greek), that is, ontologically founded on the rejection of what is Black, this term referring both to the question of race – a Black subject being, then, in the Afro-pessimist terms of Frank B. Wilderson III, "the very antithesis of a Human subject" (9) – and to outer space and its disturbing darkness.[2]

The first section will begin by analysing these three ontological reductions characteristic of the Anthropocene, after which I will show how Afrofuturism – including musicians (Sun Ra, P-Funk), writers (Alexis Pauline Gumbs, Octavia E. Butler) and a painter (Wangechi Mutu) – could help us to "recosmize the Earth," to take up a strong expression used by Augustin Berque. In this article, Afrofuturism is understood as a cultural and metaphysical alternative to white domination as it was carried out from the slave trade to the New Space. As I will show in the second section of the article, this alternative consists in confronting the technological possibilities of the Anthropocene with a *major impossible*, on the basis of which it will be a matter of imagining techniques – artistic, symbolic, anthropological – of recosmization. The cosmic techniques of Afrofuturism (third section) envisage outer space not as a space of conquest or exploitation, but as a mediation, a stellar detour thanks to which our terrestrial condition can be rethought. The Afrofuturist *cosmos* is an inter-sidereal space, not firstly a set of objects or subjects but a dark dimension, an *atopia* – a space out-of-space, a spacing – from which new ways of considering human beings and the Earth could emerge.

In relation to the *cosmos*, the Afrofuturist motto is the following: the Copernican revolution remains unfinished; in fact, it has barely even begun, given that we are still living, psychologically and politically, in Aristotle's geocentric system – in other words, the Anthropocene is cosmologically Aristotelian. The suggestion that the Copernican revolution may not yet even have taken place is in conformity with the Afrofuturist conception of temporality: in the fourth section, I will show how Afrofuturism is an upheaval of modern, white temporality – the "official, colonial, normatively white timelines" (Lothian 102) – that guide the geo-technological production of the Earth. To shed light on this upheaval, I will draw upon Walter Benjamin's philosophy and his messianism of the past, that is, the politico-theological injunction to remake the past such as it might have been if its promises of happiness and emancipation had not been annihilated by the victors of time – that is, those for whom the potential of the past must be destroyed in order that their future alone will be assured. In a similar way, Afrofuturism invents alternative futures vis-à-vis the sole future that the victors of time would like to impose by force and by datamining. The "counter-futures" of the Afrofuturists reinterpret the past in which their cosmic techniques try to disseminate, through a kind of anachronistic justice, a truth erased by the chrono-masters.[3]

It is thus in space-time as such that the cosmochronic techniques of Afrofuturism try to invest, and which they try to transform. To the technologies of the Anthropocene whose goal is to fashion an Earth 2.0, Afrofuturism opposes techniques of the imaginary that consist in the revelation of other space-times: not to remake the Earth but to dissolve it – to reduce it to zero – with a view to a transcendental accretion, a liberation of the Earth and for another *cosmos*. In a fifth and final section, I will show how the Afrofuturist "extraterrestrial turn"[4] tends to liberate the Earth and reveal an *alien*, radically non-anthropocentric universe. It is thus towards a major ontological reversal that this article is directed: to posit the alien – the zero, the Black – at the origin is to overturn the hierarchy of the Anthropocene, to make the world definitively foreign, to set a negative precondition before any technological achievement, the impossible before the possible, the mythical and the magical before the technological. In an alien universe, technologies become *Unheimliche*, definitively preventing human beings from considering technology as a way of mastering the Earth. Against the acosmic geo-technology of the Anthropocene, we must promote the cosmic techniques of the alienocene: this is what the Black Angel of the cosmic history of planet Earth has come to tell us.

## the blind spots of the anthropocene

First of all, let us shed some light on the three characteristics of the Anthropocene that I seek to question on the basis of the cosmic techniques of Afrofuturism: anthropocentrism, geocentrism and leukocentrism.

### (1) anthropocentrism

Bernard Stiegler is right to assert that a primitive name for the Anthropocene is *Gestell* – the "*arraisonnement*" (to take up the term used in French to translate Heidegger's concept) of nature, forcing it to unveil itself and to make itself available for anthropo-logical operations (agriculture, trade, war, etc.) (28). But where the effect of the concept of *Gestell* was to show that the human itself is "challenged more originally than are the energies of nature," "committed" by *Gestell*, and where Heidegger argued that "Modern technology as an ordering revealing is, then, no merely human doing" ("Question" 18–19), the Anthropocene reduces geo-technological operations to something only human, regardless of whatever narratives are proposed in order to date its origin. It is not a question of denying the political responsibility of human societies and their infatuation with a capitalism that sovereignly destroys the conditions of life, but of not forgetting – as is the case with the modern schema – that our societies are responsible for, and respond to, something that is not just human, something that they pass through, communicate and modify, without, however, being either its origin or its master.

### (2) geocentrism

The Anthropocene is incomprehensible without the *reversal of the Frontier*: on a historical level, investment, in financial terms but also in libidinal and psychic terms, which lay at the heart of the conquest of space – the Space Age – since at least the launch of the first Sputnik, was reterritorialized in the first decade of the twenty-first century from outer space back to Earth, the latter now being treated as a planet to be terraformed, even though terraforming initially meant turning another planet into one similar to Earth. In other words, geoengineering is a displaced terraforming, a terrestrialized terraformation. Terraforming Terra means having in mind an Earth 2.0 with augmented reality, having fully enlisted nature into the project of planetary re-formation. Global digitalization, which led to the modelling of the entire Earth, from DNA to climate, was the material condition that enabled the installation of the Anthropocene understood as the complete reformatting of the Earth. In other words, we have gone from the cyborg epoch of the digital era and hybrid beings, which reached its peak in the 1990s, to the neo-terrestrial epoch of the analogical resurrection of the Earth through the realization of its digital schema.

### (3) leukocentrism

Insisting on white centrism instead of white supremacy does not mean downplaying the politico-ideological manifestations of racism, but does mean making room for an examination of the ontological place occupied by the White, in order to show that this occupation comes instead and in place of what is Black so as to mask it, erase it or subjugate it. According to Frank B. Wilderson III, this indeed amounts to a foreclosure produced by the "new ontology" of modernity, an a priori rejection of Blackness in which the slave is equivalent to a figure of the "anti-Human" (11, 18). The foreclosure of Blackness, however, is also the core of the dominant cosmology: from the Space Age to New Space, it is reduced to obsessions with objects and materials, things and areas to be colonized and exploited, and one of the spearheads of astrophysics for the general public today is the question of exo-planets, reduced to those that would resemble Earth, and which would therefore be – imaginarily – colonizable. What this impoverished cosmology represses is the dark space between the planets, interplanetary and interstellar space, the overwhelming darkness that is predominant throughout the universe: observable matter constitutes only 5 per cent of the universe, to which dark matter adds 27 per cent and dark energy 68 per cent, both of which can be detected only indirectly via their effects. At the cosmological level, as at the anthropological and political levels, what is Black is repressed, ejected from the centre, scotomized or subjected to the Anthropocene and its crutch called New Space.

To leuko-, geo- and anthropo-centred geotechnology, I would, in the following sections, like to oppose Afrofuturism. But hold on! What can it mean to oppose to the real, terrible power of geoengineering an art, a speculative cosmology composed of counter-narratives (Octavia E. Butler), offbeat music (Sun Ra, P-Funk) and cosmo-human paintings (Wangechi Mutu)? It is nevertheless quite possible that today, a cosmic art offers the only real way of opposing geo-technology – an art capable of proposing another use of technology, another articulation between technical objects and the universe, or in other words, another technological individuation. This other individuation is necessary, because there where geo-technology makes the danger grow, nothing can save us, and what we need to think and promote today is instead a phase shift, a divergence, a radical opposition to the dominant technology.

## pyramids' afronauts: cosmic techniques of the impossible

Even so, is Afrofuturism capable of mounting any such radical opposition? The term Afrofuturism is said to have been coined in the early 1990s by the writer Mark Dery during an interview with Samuel R. Delany, Greg Tate and Tricia Rose. The error would consist in quoting only the beginning of Dery's definition, where he insists on the "context of twentieth-century technoculture" and the way in which African-American artists and thinkers have appropriated "images of technology and a prosthetically enhanced future."[5] Focusing only on this part of the definition risks reducing Afrofuturism to a celebration of technology, taken as the source of new and unprecedented possibilities for existence. This is how we could interpret the Afrofuturist film *Black Panther* (dir. Ryan Coogler, 2018), which imagines the secret existence of a sub-Saharan African country, Wakanda, possessing the most highly developed technology of any country in the world. Yet just after discussing the "technologically enhanced future," Dery adds:

> The notion of Afrofuturism gives rise to a troubling antinomy: Can a community whose past has been deliberately rubbed out, and whose energies have subsequently been consumed by the search for legible traces of its history, imagine possible futures? (180)

"Can the subaltern speak?," wonders Spivak, and Dery asks in his turn: can the African-American artist imagine (possible futures)? Indeed, let us not forget that the possibilities invented by the Afrofuturist authors are conditional upon a major impossible, that which put an end to the possibilities of existence of Africans in the age of slavery, this end of the world that erased forms of life, along with the names of slaves and their history, and that still continues today to destroy the lives of Black Americans, as Black Lives Matter reminds us.[6]

This place given to the impossible lies at the heart of Sun Ra's art:

> I'm talking about something that's so impossible […] it can't possibly be true. But it's the only way the world's gonna survive, this impossible thing […] My job is to change five billion people to something else. Totally impossible […] Everyone else can do the possible things, why should I waste my time with that? Everything that's possible's been done by man; I have to deal with the impossible.[7] (Corbett 311–13)

Here we can see the stark difference from the geo-technology of the Anthropocene, which, faithful to the modern categorical imperative, postulates that nothing is impossible, that everything can and even must be produced, realized, from nuclear power to fracking via geoengineering and GMOs (genetically modified organisms). By contrast, Afrofuturism develops a *technique of the impossible*: the revelation of the impossible at the very heart of technological possibility. In the Afrofuturist imagination, this is precisely what the pyramid symbolizes. In the P-Funk song "Mothership Connection (Star Child)," from the album *Mothership Connection* (1975), we hear the lines: "Starchild / Citizens of the universe, recording angels / We have returned to claim the pyramids," and, in the "Prelude" of the album *The Clones of Dr. Funkenstein* (1976):

> In the days of the Funkapus, the concept of specially-designed Afronauts capable of funkatizing galaxies was first laid on Manchild but was later repossessed and placed among the secrets of the pyramids, until a more positive attitude towards this most sacred phenomenon – clonefunk – could be acquired. (Eshun, *More Brilliant* 142)

Unlike *Black Panther*, where Wakanda's high technology is supposed to have existed without interruption and long before the development of "Western" scientific modernity, here funk technology would have been deactivated and placed in the pyramids: "to claim the pyramids" thus means to revive [*réactiver*] a technology from the past, to bring it back to life (hence the name Dr. Funkenstein etc.). Revival, of course, is a form of relation that lies at the heart of the technical-imaginary function of the pyramids: to preserve bodies and prepare them for the afterlife. But this function, which brings us into the presence of the impossible as such – the impossible resurrection of bodies, that impossible whose name is death – is first and foremost the manifestation of the essence of the pyramids: to be a cosmic machine, that is, having at stake the relationship between the Earth and the universe. Indeed, what requires investigation is the cosmological topology of the pyramid.[8] The mummified body is placed in the substructure of the pyramid, underground, where it is to be preserved, intact, for eternity. And in the superstructure, the tomb itself that is located on the surface, food and drinks are arranged, for the body that will supposedly need them in the afterlife. Hence the pyramid establishes a relationship between a subsoil-eternity and a surface-based-afterlife, as if it were a question of finding the way of passing from one to the other, from the mummy's desiccated interior to the outside, where life is revived. This is because pyramids involve both a horizontal terrestrial axis, joining a river to the desert (in other words, creating a link within the fundamentally bipolar geographical structure of Egypt) through a series of bridges and corridors, and a vertical axis pre-formed by the subsoil/surface (burial/tomb) relationship. For the pyramid has the ultimate imaginary and symbolic function of being a "place for Ascension," a "bridge" between our world and the next, between the Earth and the Sun, the terrestrial plane and the extraterrestrial plane. Could we not therefore consider the pyramid to be an instrument of cosmic communication, allowing the symbolic exchange Earth/Sky – or Earth/Sun – and making it possible for the deceased king to pass from the Earth to the solar eternity and to return from it back to Earth if necessary? Is it not this symbolic exchange between Earth and universe that Afrofuturism reclaims and seeks to revive?

## deep sea, black icarus, dark space

While the Sputniks may have begun their circuits around the Earth in the early 1950s, the so-called Cold War being measured by the amount of energy burned by the rockets of Russia and the United States, Sun Ra proposes another Space Age – "Space is the place," he tells us. For the Afrofuturists, space is first and foremost the promise of other possible worlds: as the title of one of the songs of Sun Ra puts it, "There are Other Worlds (They Have Not Told You Of)" (1978). These other worlds are promised to those who, having endured slavery, endured, on Earth, the end of their world. And it is this terribly missing world that the Afrofuturists seek elsewhere:

"if we came from nowhere here / why can't we go somewhere there?" (Ra 206). Here, on Earth, is nowhere, for those who have lost their world:

> This world / Is not my home. / My home is / Somewhere better. / Many light years in space, / My home. / In some far place / In light years in space / Where human beings have never run. / Where your eyes have never seen. / My home, my home / planet?? (Ra 381)

As the title of a Sun Ra song from the early 1980s says: "They Plan to Leave." But here again we need to underline the *dimension of the impossible* that the possibilities offered by aerospace engineering could lead us to forget, the impossible that haunts the voyage through outer space promoted by the Afrofuturists:

(1) On the one hand, traversing outer space repeats another crossing, that of the Atlantic Ocean in a triangle of traffic – a crossing far from the familiar world, a perilous crossing far from land. In her book, *M Archive: After the End of the World*, the poet, activist and independent scholar Alexis Pauline Gumbs – inspired by the Afrofuturist tradition and the poet Audre Lorde – insists on the homology between ocean and universe:

> they had this thing about darkness. the bottom of the ocean, outer space. they were afraid of it, they wanted to penetrate it. they wanted to pretend it only existed in contrast to light. and there was something about now wanting to have their eyes closed. not wanting to go within. lightweight enlightenment equaled mass black death. (Gumbs 10)

In this remarkable passage that picks up on what I call leukocentrism (a darkness supposedly existing only "in contrast to light") and adds a gender dimension ("penetration" with a destructive purpose), Gumbs marks with a simple apposition (a comma articulating "the bottom of the ocean" and "outer space") the correspondence between deep ocean and infinite space, a correspondence that works thanks to the mediating presence of darkness. This correspondence, which the Afrofuturist tradition constantly highlights, also makes it possible to identify slave ships and spaceships. Finally, it is worth noting that this correspondence gives rise to an extraterrestrial imaginary, concerned with what comes from beyond space (the space capsule on the cover of *Mothership Connection* and the flying saucer that would land during P-Funk concerts in the 1970s), and the invention of an ocean mythology: Detroit's famous electronic music duo, Drexciya, imagined the existence of the Drexciyans, beings descended from pregnant black slaves thrown overboard during the Atlantic crossing, but who survived, mutated and became amphibious.[9]

(2) On the other hand, the voyage through space repeats another aspect of the Middle Passage since the hypothesis of flying away to escape a situation of servitude does not date from the Space Age. It dates from slavery and the plantations, and it was expressed in the form of tales of "flying Africans" who, possessing the magical power of flight, were able to escape from the plantations and return to Africa. According to Terri L. Snyder, these tales are in fact a retroactive, "transfigured" version of slave suicides, or in other words, a fictional way of "redressing the wrongs of history" (14, 157–66). The Black Icarus that the Afrofuturists offer us is drawn to the Sun only to be – inversely and simultaneously – weighed down by a dark and irreducible kernel of self-destruction, amounting to a plan to leave when there is no other plan. And if the flight of Africans out of their condition as slaves is a rewriting of a tragic past for the sake of a future, then Afrofuturism and its aspiration of intergalactic flight is the rewriting of this rewriting, its resumption, because the transfiguration has failed, at least partially, and it will perhaps always fail, thanks to the impossible kernel of self-destruction that inhabits it.

So it is not only that "they plan to leave" in a universe that would remain unchanged, it's also that "they" plan to resymbolize the whole universe and the relation between planet Earth and the other planetary or stellar bodies: other worlds are not spaces to be discovered and conquered, but worlds that will arise from a reconfiguration of the universe. Sun Ra – as his name suggests – would therefore be the embodiment of a Copernican revolution that would finally take place, that would finally make the Sun prevail on Earth. "The isolated Earth age is finished," Sun Ra announces (307), and "the galaxies are waiting / For planet Earth's awakening" (331). This desire for a Copernican revolution is illustrated by the album cover for *The Heliocentric Worlds of Sun Ra* (1965), where we see the figure of Sun Ra set among those of Copernicus, Galileo Galilei and Tycho Brahe.

But in the Copernican revolution proposed by Afrofuturism, the sun is not the only centre, and the Afrofuturist universe is at least bipolar: there is the Sun, but there is also radical darkness, the bottomless Black of the ocean floor and outer space; there is this end of the world that comes from the Middle Passage and that is coming again, as Alexis Pauline Gumbs tells us, with the ecocide now underway. Sun Ra was well aware of this, being one who – while privileging the Sun – asks us to return to zero, this zero that he called the "no point": "out of nowhere they come from the no point" (257). This is what technophiles should never forget: "at first nothing is" (259). And the nothing, the negative, persists in the form of a radical darkness: the "outer darkness," the "Void-Blackness," the "Black-Infinity," the "Dark Unknown Eternal," and so on. And it is of this "blackness" that Sun Ra wishes to give an account in his "intergalactic music," a blackness "that the world does not know," "that the world will never understand" (296).

Is the Afrofuturist universe really bipolar, then, divided between the Sun on one side and darkness on the other? I prefer to see it in terms of the proposal made by the contemporary Russian painter and writer Pavel Pepperstein, whose paintings and stories postulate the existence of a cold, dark centre within the Sun itself, an eternal "astral darkness" that would be a "condition of light" (31).[10] This is all the more interesting because, in a gesture opposed to both the geo-technology of the Anthropocene and Russian cosmism and its appetite for cosmic geoengineering, Pepperstein belongs to a theo-philosophical tradition that has made its way from Malevich's *Black Square*, a tradition that consists in promoting the mystical virtues of darkness.[11] This allows Pepperstein to oppose the Russian "cosmonauts" – interested in the dark "background" of the universe, fascinated by "darkness and mystery," not by the stars but by "the dark gaps between them" – to those "astronauts" (Americans) who are fascinated simply by the "stars," which Pepperstein interprets as the expression of an individualistic ideology ("deified individuals") (29).

Far from distancing us from the Afrofuturist question, this short detour by way of Pepperstein brings us back to it since his narrative serves as a reminder of how Russia could be considered a kind of "Africa in the snow" (30). For the cosmological revolution called for by Afrofuturism does not (only) consist in decentring the Earth, but in rethinking the fundamental polarities of the (oceanic) inside and the (galactic) outside, the polarities of Earth and the universe: everything communicates through darkness, this being the fundamental dimension of the universe. Black Icarus voyages between the stars, through the darkness that joins them by separating them, and makes communication possible through the Middle Passage and the passage to the stars, Africa and North America, the past and the future. All of these terms can be exchanged with one another in order to create a place, a moment, a new and unprecedented present. So, what then is space? A place for a radical reorganization of the universe through its dark dimension, which acts as an ex-change, an inter-change. Rather than leaving Earth, it is ultimately a question of drawing this Earth into this voyage into the unknown: "You're on the spaceship Earth. Destination / unknown" (Ra 443).

## future conditional and the copernican revolution of temporality

> This is the space-age [...] / The age beyond the earth-age / A new direction / Beyond the gravitation of the past. (Ra 313)

In Afrofuturism, it is always a question of thinking both space and time. In the above quotation, the past is associated with the gravitation from which Black Icarus must tear himself away, and, in another poem, Sun Ra plays on the signifiers grave and gravity, connecting the gravitation of the Earth to the death produced by slavery.[12] But what I showed earlier amounts to the following: leaving the Earth in no way means leaving the kernel of negativity, because Black Icarus takes gravity with him wherever he goes. And the future, as Sun Ra knows very well, will never be free of this original negativity: "the future is never / Never comes tomorrow / Never is not" (252). "Never is not" does not mean that "never" is necessarily negative, but that the future is not the abolition of the past's negativity: the future must constitute another relation with negativity, with the astral darkness that even the Sun contains.

This is the key Afrofuturist lesson concerning time, its inaugural paradox: we must invent the future even when there is no longer any possible future. The Afrofuturist future must be written for those who are in a position of "structural futurelessness" (Lothian 100), even though the slaves were not supposed to survive. This is what the poet Audre Lorde writes in her famous poem, "A Litany for Survival," about those who "were never meant to survive" (255). This survival in spite of everything concerns the Black condition, from slavery to the contemporary situation of Black Americans who suffer a "social death" or a death at the hands of police violence. But then, what kinds of futures are described in Afrofuturist science fiction texts, and what is their relationship to the past, if the past amounts to the prior destruction of any possible future?

The hypothesis I am defending is that the futures of Afrofuturism seek to remake the past – in Walter Benjamin's sense. In the notes of the *Arcades Project*, Benjamin refers to a "Copernican revolution" that would affect historical perception: what we used to think is that the past is "a fixed point" – just as the Earth used to be conceived as a fixed point within the geocentric perspective – and that the present can be modified, depending on our capacity to produce a knowledge on the "ground" of a past that is "fixed" (388). But "now this relation is to be overturned, and what has been is to become the dialectical reversal – the flash of awakened consciousness" (388). What Benjamin calls "the dialectical, Copernican turn of remembrance" (389) is a "form of remembrance" that is able to change the past: "what science has 'determined,' remembrance can modify [...] Such mindfulness can make the incomplete (happiness) into something complete, and the complete (suffering) into something incomplete" (471). This point is fundamental: to declare history complete, finished, is necessarily to declare the victors definitively victorious, it is to consider that the hopes of liberation of the vanquished of the past have been irretrievably annihilated by the oppressors, and it is thus to render futile any idea of future justice. In its most vivid works, however, Afrofuturism consists in reopening the past, thanks to the future.

Reopening the past can be, as Kodwo Eshun writes, a form of "revision" that, for example, consists – Eshun refers to Martin Bernal (*Black Athena*, 1988) and George G.M. James (*Stolen Legacy*, 1989) – in exposing "the white conspiracies that covered up the stolen legacy of African science" ("Further Consideration" 296–97), or – with Toni Morrison – that African slaves were the first modern subjects, even "post-modern," in that they, and especially they, were the first subjects to experience the "dissolution" of their world, a radical ontological instability, a form of madness tied to the dehumanization produced by slavery (Gilroy 178).[13] But what Benjamin insists upon are not the realities of the past that have been occluded so much as forms of unexpressed, unrealized potentialities: what is to be remembered is the dream that bore the lived reality.

And the goal of what would amount to a political relationship to history understood starting from its promises is "to ignite the explosive materials that are latent in what has been" (Benjamin 392). In other words, the archive that Benjamin asks us to explore is one that has not been written, or insufficiently written: "to read what was never written," in the words that Benjamin takes from Hofmannsthal, is the very task of the historian (Benjamin 416). Reparative writing, then, does not consist in restoring the historical truth of what took place, but in establishing historical justice for what did not take place. Not: what would you dream had happened in the past? But: what dream in the past happened to you so that you should remember it?

Let us add, however, that in the Anthropocene epoch, the apocalypse is redoubled: it no longer comes just from the past but also from the future; it no longer concerns only the vanquished of a history that has taken place but also those of history to come, that of the ongoing ecocide, of the Sixth Extinction – of the environmental apocalypse. In *M Archive*, Alexis Pauline Gumbs pursues the Afrofuturist line out into the twenty-first century: survival, for those who are destined not to survive because of the death drive that governs the Anthropocene, now takes place between two apocalypses, that of the past and that of the future. But then, one might say, what does it mean to imagine futures if, all of us, all become "structurally futureless"? It means: to invent conditional futures. The conditional past expresses itself thus: if we had done that, in the past, then things would have been different. But what I call future conditional is the effect of the translation – in a geometric sense – of the past into the future: if the future had existed, in what is already thinkable as a non-future, then it might have been this one, the one that this or that Afrofuturist writer tells us about.

In this situation of being between-two-apocalypses, we are already survivors who are living the absolute intensity of a present loaded with all kinds of extinctions, those of the past as well as the future. Never has Ytasha L. Womack's phrase – "Afrofuturism is where the past and future meet" – taken on such force as it does today, and in such dramatic fashion. In fact, everything happens as if the past and future were now flowing back into the present, seeking refuge there. In this sense, the present now bears every temporality within it, superimposed upon it: it must encounter the dreams of both the past and the future, and in turn produce its own way of dreaming anew, for its own sake. We are, therefore, far from a linear, white temporality oriented by the idea of progress. The temporal figure that emerges, then, is that proposed by Michelle M. Wright in *Physics of Blackness*, where she analyses Blackness not as a *what*, for which it would be a matter of determining an essential cause lying in the past, but as a *when*, which would be thinkable on the basis of an "intersection" of social constructions at a precise moment of time, no longer a straight line but "a circle with many arrows pointing outward in all directions": "the circle denotes the 'now' of the present moment, and the arrows represent all the spacetimes that intersect with that 'now'" (20). A circle with arrows pointing in all directions? We will have little trouble recognizing the figure, Afrofuturist if ever there was one, of a sun – a sun radiant with sacrificed temporalities.

## cosmos alien

The cosmic techniques of Afrofuturism aim at an upheaval of space-time. Far from being repressed in favour of the planets, the interstellar Black, this atopian environment, becomes a vector of communication. And the present – which has become an achronic dimension – becomes the bearer of both the past and the future. This radical change in the markers of space and time is constitutive of the *alien* universe that Afrofuturism has in its sights. As Mark Sinker showed in a pioneering article published in 1992, the apocalypse of slavery is like an equivalent to alien abduction, with colonizers being comparable to extraterrestrials.[14] This comparison is already at work in *The Wretched of the Earth*, where Franz Fanon analyses colonialism as a situation in which not two classes but two different "species" are produced:

the ruling species is first and foremost *the outsider from elsewhere* [my emphasis], different from the indigenous population, "*the others*" [my emphasis]. (5)

On one side, then, there is the alien colonizer, and, on the other side, another alterity, that of "the others": at the heart of Afrofuturism reigns this ambivalence of the alien that is dangerous and destructive but also a vector of beneficial transformation. We find this ambivalence in Octavia Butler's famous trilogy, *Lilith's Brood*, previously collected under the title *Xenogenesis*. In this trilogy, an extraterrestrial race saves the human species from complete extinction only by forcing it to share its genes with it, an exchange that leads to mutations for both species – humans becoming (or enveloping) an "alien Humanity" (*Lilith's Brood* 449). The terrestrials must exchange, must negotiate their own identity, with the extraterrestrials, compelled to "make some kind of accommodation," as Butler writes in her afterword to her short story "Bloodchild" (*Bloodchild* 32). In this sense, Butler's trilogy is the best possible illustration of a statement by Eshun that could be taken as pinpointing the essence of Afrofuturism: "Away from alienation. Into the arms of the alien" (*More Brilliant* 157).

Let us try to visualize the type of cosmos that emerges from this radical alienation – where the term "alienation" is understood not in the Marxist sense of being dispossessed of what belongs to us and ultimately of our own essence, as dwelling in a world that has become foreign even though it is we who have produced this world, but as becoming other than what has alienated us, other than what has dispossessed us. First and foremost, it will be a question of promoting ways of becoming other-than-human: Afrofuturism does not seek to take over the category of "human," because the latter was defined, with slavery, in opposition to Blacks, but rather to found one or more other categories. In this sense, becoming-alien is parallel to the process that consists in leaving Earth, for both cases involve a refusal of the humanist, "liberal," reformist perspective, which would dream of a more welcoming Earth and concept of "humanity." We can see these alien-becomings at work in the paintings of the contemporary artist Wangechi Mutu. Composed of thirteen figures, her collage installation *Family Tree* (2012) proposes a new cosmogony deriving from two main figures, "Original sky" and "Original land" (Schoonmaker 21–50).[15] All these figures are both futuristic and traditional, global and anchored in specific places, borrowing from sci-fi, African and Asian cultures. It would be easy to see these figures as post-human or cyborgian, but it seems to me that these now stereotypical concepts in fact prevent us from perceiving the excess at play in Mutu's Afrofuturist extra-human projections.

Let us take a look at Wangechi Mutu's obviously Afrofuturist-influenced diptych entitled *Yo Mama* (2003): on a partial planetary surface, a woman with a white face and prominent lips is crushing the decapitated head of a snake with her heel. This Eve, having defeated the serpent, rewrites Biblical history, human history, which, far beyond the Christian Bible alone, has too often relied on the original condemnation of women. Yet this tableau, with its feminist dimension, does not oppose phallocentrism in order to replace it with the sovereignty of women: it should be noted that the right leg of the female figure is amputated, thus marking her incompleteness, her wounded sovereignty. The female subject and her bit of planet lie at the heart of a pinky-purple universe seemingly subject to wave motions (verifying the universe/ocean association to which I previously referred). This universe contains black suns with tentacle rays, artificial globes and planets whose physiognomy is more conventional. The beheaded snake continues into the second part of the diptych, on a planet that bears a strong resemblance to a jellyfish. This rewritten, re-coloured universe is not just an external container, but also seems to inhabit the body of the main figure, subject to nebulous, galactic motifs, turning the skin into a *cosmos* – unless it is a disease: just as the new sovereignty of a non-phallocentric universe makes women into beings who are less than omnipotent, just as a part of the black face of the "Original land" is that of a white, so too the cosmic skin is not the manic celebration of a new universe, but a surface where beauty can at each and every moment be exchanged with horror.

The *alien* cosmos of Afrofuturism allows us to reconsider the relationship of the Earth to the world, to take up the terms of Heidegger's analysis. We know that the latter, in "The Origin of the Work of Art," expressly rejects the idea that the Earth can have any relation with the universe: the Earth, Heidegger argues, has nothing to do with "the merely astronomical idea of a planet" ("Origin" 21). In this type of geocentric rejection, which Heidegger shares with Arendt and Husserl,[16] a spatial abdication is played out in the Anthropocene and its reduction to the terrestrial field. Now, I believe that it is possible to affirm that:

(1) This fault/rift that Heidegger locates between the Earth and the world could be interpreted – against Heidegger – as the dark presence of this Blackness that the Afrofuturists identify as inter-sidereal space. To think of the darkness of the universe as part of our planet is to understand how much the extraterrestrial dimension inhabits terrestriality. The Black is at the heart of the world, or rather: something Black prevents the world – of culture, art, institutions, thought – from encompassing the Earth, prevents these from being superimposed without remainder. It also means that if the Earth is at least partly indiscernible, impossible to disclose (and, in my vocabulary, I would say unconstructable), it is because the Earth belongs to the universe, because the Earth is itself the consequence of the history of the universe that began some 13.8 billion years ago.

(2) Once it is understood that Earth and world are unthinkable without the universe, it is the Heideggerian fourfold – earth, sky, mortals and divinities – that demands to be revisited from top to bottom. In the cosmogenesis that I propose, each existing thing is a *situated-alien-formation*, that is, involved in a situation where the *alien*, understood as a dark transcendental definitively disadjusting world and Earth, may be given in the transitory, non-alien form of a human or (possibly) of an immortal, or else of a plant, or an animal or a stone. Earth and sky are then the two cosmological coordinates from which the alien forms a – transitory, contradictory – non-alien identity. The terrestrials that we are, then, are in fact dialectically extraterrestrial, in the sense that we are composed of "star dust" (the Earth having been formed from the dust of supernovas).

The *alien* cosmos is therefore ultimately an alienocene: what cosmological art reveals is the existence of an alienocene at the heart of the Anthropocene, an alienocene that comes from the Black, that refutes leukocentrism in the name of the Black, anthropocentrism in the name of the *alien* and geocentrism in the name of the *cosmos*. Far from being the opposite of the Anthropocene, the alienocene is what, within the Anthropocene, opens up its internal abyss, its repressed transcendental foundation; in more positive terms, it is the expression of *alien* power and the possibility, thanks to this power, of re-founding a cosmos. In this sense, I agree with Yuk Hui's thesis: it is indeed "a new world history" that we must think (*Question Concerning* 202) and a "cosmotechnical event" that is required to neutralize the disastrous effects of the Anthropocene.[17] Hui rightly invites us to rethink together – as I tried to do for Afrofuturism – "technics, mythology, and cosmology" (*Question Concerning* 29). Nevertheless, I do not think that the conceptual grasp of the relationship between *cosmos* and technics, which Hui calls "cosmotechnics," could and should give rise to "the unification between the cosmic order and the moral order through technical activities" (*Question Concerning* 19). Firstly, the cosmos I am talking about is not reducible to the terrestrial world and its history: the cosmos traverses the entire history of the universe. Secondly, the *alien* cosmos is not so much an "order" as a counter-order, an objection to any identity and any harmony. Thirdly, this *cosmos* innervates less a morality than a politics resistant to any order: the an-archist conflagration of the universe (the so-called Big Bang) resonates in the political matter of human societies. Technics or technique would then be the echo of this conflagration, as a reminder of that impossible-and-yet-there that is the universe, and the cosmic techniques of Afrofuturism are magical less in the sense that they unify the mythical and the technological, and more because they radically transform their meaning and function (Hui, *Question Concerning* 22–23): the mythical could then become a cosmo-technological programme in search of a politics.

## black angel (cosmos and communism)

If Afrofuturism is so crucial in the theoretical approach that I am proposing here, it is because the Black can now be thought both as an interstellar milieu and as the object of (cosmological and political) repression. But the Black, as the object – and effect – of leukocentric rejection, must be ontologically overturned, not so as to be mastered and changed into a foundation, but in order to keep the *alien* power operational, a power that is unknown as such.[18] It seems to me that there is a common ontological point shared between Afrofuturism and Afro-pessimism: in both cases, it is no longer the White that is posited first, as possibility (of development, colonization, exploitation), and the Black second as non-White (the non-White produced by the White); instead, it is the Black posited first, as impossible. The White is now only the non-Black possible, a secondarized possible, and in this sense limited (Sexton 6).[19] This common point is also political: just as Afrofuturism demands a true Copernican revolution (Sun Ra) and a new cosmogony (Wangechi Mutu), Afro-pessimism demands a Black revolution, which has not taken place – except in Haiti – and which cannot be replaced by making "inclusive" social changes to North American or British liberal society.

But isn't that asking too much? A revolution, and on top of that a cosmological one? Isn't this to wander into the realms of the imagination? One might respond to this as Sun Ra does in the film *Space is the Place*, when young people make fun of his simultaneously archaic and futuristic attire, telling him that it doesn't look "real." He replies:

> I'm not real – just like you. You don't exist in this society. If you did, your people wouldn't be seeking equal rights [...] So we're both myths. I do not come to you as a reality; I come to you as the myth. Because that's what black people are: myths. I came from a dream that black man dreamed long ago. I'm actually a presence sent to you from your ancestors.[20] (Youngquist 220–21)

For Sun Ra, the myth is not something belonging to the past, it is the inscription in the present moment of what lacks political existence. It is the atopia of the *polis* and the achrony of politics – their darkness and their zero. It is also, in Benjaminian terms, the "dream" of the past that has not been realized. This is what the Black Angel of History, in its Afrofuturist variation, comes to tell us: the unjust mythical unreality must find a real resolution, so that the absence of future – the zero point of the coming time – becomes "a point from which radical disruptions can be enacted" (Lothian 115). To speak of the Black Angel of History is, of course, to refer to one of Benjamin's famous "theses on history," where the Angel represents the being who would like to save humanity from the disasters of progress, but is unable to do so because it is irresistibly pushed forward into the future. But it is also to refer to the fiction-documentary film, *The Last Angel of History* (dir. John Akomfrah, 1995), in which a character named Data Thief comes from the future to find – thanks to his crossing of the "black hole" of history – a space-time in which past and future combine, and abolish the linearity of White time oriented by "progress."[21]

Now, I believe it is possible to draw from these analyses elements of an answer concerning the political and cosmological function of the Black Angel of History, and to do so by recalling that Sun Ra claimed to come from an "angel race" (or from Saturn, the two not really being different for him) (381). The Black Angel is in the position of mediator, or rather, of medium, because of his Black (non)colour, which is that of the unknown, and because of the loss (of colour, of particularity, of way of life, of habit) to which any radical subjective and collective transformation gives rise. The Black Angel is thus always in the middle [*milieu*] – interstellar – and in this sense always the medium [*milieu*] par excellence, that within which any object of the universe gravitates. In temporal terms, the Black Angel is the present onto which flows back, in the Afrofuturist perspective, the past and the future subjected to their respective apocalypses. It is in this sense an ark, to evoke here the name of Sun Ra's orchestra (the Arkestra), or a spacecraft – except that it is not a question of leaving the Earth, but of changing the symbolic place of the latter. In the same way, it is not a matter of leaving the present for the future or for an imaginary past (as in the film *Black Panther*), and nor is it a matter of repairing the damage of history, but rather of changing the present in order to transform it into a place of the fusion of the political promises of emancipation and happiness, whether they come from the past or future.

The Black Angel is thus opposed to the White Machine: in a battle of titans opposing the myth of Passage – through history towards the present and its crossroads where multiple temporalities conjoin – and the myth of the geo-technological Frontier. The mythico-poetic techniques of Afrofuturism aim to reveal this *cosmos*, radiating from intense presents, having collected past and future, and the technologies that these techniques utilize are no longer instruments of power – of White on Black, of the Machine on the Earth or some asteroid – but revelations of the impossible. In a *cosmos* become *alien*, the mega-machine of control changes into the para-machine of liberation. Technologies then express the strangeness of the world: they no longer serve to accomplish the possessive will of a nation or an individual, but they express an inter-relation – black, obscure, going towards zero, atopian – that belongs to no one. To no one, or perhaps I should say to everyone: the *logos* of this *tekhne* concerns justice, not exploitation, beauty, not utility, truth, not efficiency – it signals towards a *cosmunism*.

## disclosure statement

No potential conflict of interest was reported by the author.

## notes

1 I thus maintain that what Koyré described as the passage from the "closed world" to the "infinite universe" did not simply lead to the end of all *cosmos*, but to the necessity of redefining this term, neither as order nor as pure chaos, but as the *historical formation of the alien-totality*.

2 Even neurology can be leukocentric: see Fern.

3 On the concept of "counter-futures," see Eshun ("Further Consideration" 301).

4 A turn that, according to Kodwo Eshun, enables "an identification" with "the potentiality of space and distance within the high-pressure zone of perpetual racial hospitality" ("Further Consideration" 298–99).

5 It is this part of Dery's definition that, for instance, Reynaldo Anderson and Charles E. Jones quote in their co-edited volume *Afrofuturism 2.0: The Rise of Astro-Blackness* (viii).

6 See what Saidiya Hartman says about "the impossibility of fully recovering the experience of the enslaved and the emancipated" (10).

7 Sun Ra, "Gravity and Levity" (Corbett 311–13).

8 Here I am relying on David.

9 We can read this myth in the sleeve notes for their CD, *The Quest* (1997).

10 See also the painting, *Expedition into the Sun* (Pepperstein 26).

11 On Russian cosmism, see Groys.

12 See also: "Are you a slave to planet Earth? You said you're free, Prove it to me" (Ra 21).

13 Toni Morrison interviewed by Paul Gilroy (Gilroy 178).

14 On this point, see Sinker.

15 I rely here on Trevor Schoonmaker (21–50).

16 After all, didn't Husserl maintain that "the Earth does not move"?

17 See Yuk Hui, "For a Cosmotechnical Event" and my response to Hui's essay in the same issue: "Cosmos and Technology (*Dasein*'s Planetary Condition)."

18 I will thus not try here to show – borrowing the terms of Achille Mbembe – how the "Negro" is "produced" (68) and how the "*becoming-black of the world*" (17) is put in place, not by rejecting this analysis, but because my objective in this article is to measure the effects of a reversal: the Black as a cause and not as a product; as returning [*revenir*] and not as becoming [*devenir*]. Mbembe certainly conceives a reversal of the negative meaning of the word "negro" into a positive meaning: no longer the rejected, fungible, invisible being, but the possibility of a larger humanity, going beyond races and differences, a world-community (230, 257–63). But this reversal of "nothing" into "everything," which is not so far from what Sieyès said about the Third Estate or Marx about the proletariat, does not seem conceivable to me without first acknowledging to what extent the world is not one: it is divided between the world of capital and the world of those who refuse it (on this point, see Kisukidi 98–99). It is on the basis of this division that the Afro-pessimist thesis seems relevant to me: its reversal of the meaning of the word Black leads to a political division and not to a cosmopolitan unification.

19 I lean here on Jared Sexton's article (6).

20 Sun Ra, quoted in Youngquist 220–21.

21 I lean here on the interpretation given by Clinton Fluker, via the analyses of Houston Baker devoted, in his book *Blues, Ideology, and Afro-American Literature*, to the meaning of this crossroads in the blues (see Fluker).

## bibliography

Anderson, Reynaldo, and Charles E. Jones, eds. *Afrofuturism 2.0: The Rise of Astro-Blackness*. Lanham: Lexington, 2016. Print.

Baker, Houston A., Jr. *Blues, Ideology, and Afro-American Literature*. Chicago: U of Chicago P, 1987. Print.

Benjamin, Walter. *The Arcades Project*. Trans. Howard Eiland. Cambridge, MA: Harvard UP, 1999. Print.

Berque, Augustin. *Recosmiser la Terre. Quelques leçons péruviennes*. Paris: Éditions B2, 2018. Print.

Butler, Octavia E. *Bloodchild*. New York: Seven Stories, 2005. Print.

Butler, Octavia E. *Lilith's Brood*. New York: Grand Central, 2007. Print.

Corbett, John. *Extended Play: Sounding Off from John Cage to Dr. Funkenstein*. Durham, NC: Duke UP, 2014. Print.

David, Rosalie. *Cult of the Sun: Myth and Magic in Ancient Egypt*. New York: Barnes, 1998. Print.

Dery, Mark. "Black to the Future: Interviews with Samuel R. Delany, Greg Tate, and Tricia Rose." *Flame Wars: The Discourse of Cyberculture*. Ed. Mark Dery. Durham, NC: Duke UP, 1994. Print.

Drexciya. *The Quest*. Submerge, 1997. CD.

Eshun, Kodwo. "Further Consideration on Afrofuturism." *CR: The New Centennial Review* 3.2 (2003): 287–302. Print.

Eshun, Kodwo. *More Brilliant Than the Sun*. London: Quartet, 1998. Print.

Fanon, Frantz. *The Wretched of the Earth*. Trans. Richard Philcox. New York: Grove, 2004. Print.

Fern, Robert. "The Leukocentric Theory of Neurological Disorder: A Manifesto." *Neurochemical Research* 42 (2017): 2666–72. Print.

Fluker, Clinton. "Akomfrah's Angel of History." *liquid blackness* 1.4 (Nov. 2014). Web. <http://liquidblackness.com/issues/LB4essays/LB4CF.pdf>.

Gilroy, Paul. "Living Memory: A Meeting with Toni Morrison." *Small Acts: Thoughts on the Politics of Black Culture*. London: Serpent's Tail, 1993. Print.

Groys, Boris. *Russian Cosmism*. Cambridge, MA: MIT P, 2018. Print.

Gumbs, Alexis Pauline. *M Archive: After the End of the World*. Durham, NC: Duke UP, 2018. Print.

Hartman, Saidiya. *Scenes of Subjection: Terror, Slavery, and Self-Making in Nineteenth-Century America*. Oxford: Oxford UP, 1997. Print.

Heidegger, Martin. "The Origin of the Work of Art." *Off the Beaten Track*. Trans. Julian Young and Kenneth Haynes. Cambridge: Cambridge UP, 2002. Print.

Heidegger, Martin. "The Question Concerning Technology." *The Question Concerning Technology and Other Essays*. Trans. William Lovitt. New York: Garland, 1977. 18–19. Print.

Hui, Yuk. "For a Cosmotechnical Event." *Foundations of Science*. Forthcoming.

Hui, Yuk. *The Question Concerning Technology in China. An Essay in Cosmotechnics*. Cambridge: Urbanomic, 2016. Print.

Kisukidi, Nadia Yala. "Le nom 'Noir' et son double." *Esprit* 12 (2018). Print.

Koyré, Alexandre. *From the Closed World to the Infinite Universe*. Baltimore: Johns Hopkins UP, 1957. Print.

Lorde, Audre. "A Litany for Survival." *The Collected Poems*. New York: Norton, 1997. Print.

Lothian, Alexis. *Old Futures: Speculative Fiction and Queer Possibility*. New York: New York UP, 2018. Print.

Mbembe, Achille. *Critique de la raison negre*. Paris: La Découverte, 2013. Print.

Neyrat, Frédéric. "Cosmos and Technology (*Dasein*'s Planetary Condition)." *Foundations of Science*. Forthcoming.

Pepperstein, Pavel. "The Black Star." *The Cold Center of the Sun: Short Stories*. Ostfildern: Hatje Cantz, 2015. Print.

Schoonmaker, Trevor. "A Fantastic Journey." *Wangechi Mutu: A Fantastic Journey*. Ed. Trevor Schoonmaker. Durham, NC: Nasher Museum of Art, Duke U, 2013. Print.

Sexton, Jared. "Afro-Pessimism: The Unclear Word." *Rhizomes: Cultural Studies in Emerging Knowledge* 29 (2016). Web.

Sinker, Mark. "Loving the Alien – In Advance of the Landing." *The Wire* 96 (Feb. 1992). Web.

Snyder, Terri L. *The Power to Die: Slavery and Suicide in British North America*. Chicago: U of Chicago P, 2015. Print.

Stiegler, Bernard. *Qu'appelle-t-on panser? 1. L'immense régression*. Paris: Les Liens qui Libèrent, 2018. Print.

Sun Ra. *The Immeasurable Equation – The Collected Poetry and Prose*. Ed. James L. Wolf and Hartmut Geerken. Wartaweil: Waitawhile, 2005. Print.

Wilderson, Frank B., III. *Red, White and Black: Cinema and the Structure of U.S. Antagonisms*. Durham, NC: Duke UP, 2010. Print.

Womack, Ytasha L. "Ytasha Womack on Afrofuturism and the World of Black Sci-Fi and Fantasy." *Bitchmedia* (18 Nov. 2013). Web. <https://www.bitchmedia.org/post/interview-with-ytasha-womack-on-afrofuturism-and-the-world-of-black-sci-fi-and-fantasy>.

Wright, Michelle M. *Physics of Blackness: Beyond the Middle Passage Epistemology*. Minneapolis: U of Minnesota P, 2015. Print.

Youngquist, Paul. *A Pure Solar World: Sun Ra and the Birth of Afrofuturism*. Austin: U of Texas P, 2016. Print.

peter skafish

# EQUIVOCATIONS OF THE BODY AND COSMIC ARTS

## *an experiment in polyrealism*

The stories are legend, yet not all of them are fantastic. In China, adepts of *qi gong* still cultivate wisdom and extraordinary physical abilities by "breathing" through their pores, while others in their lineages use related techniques to remain lucid and vital at advanced ages. Then there are the doctors of the country's less occult but still ancient medical tradition, who succeed at anesthesia and treating major disease by the same cosmological principles as their Daoist counterparts. Perhaps they cure only despite those florid notions, but they may be right to attribute their efficacy to working with *yinyang* and *qi* (氣).[1]

It was long inconceivable that the question of what a body can do concerns such corporeal arts, but the impasse to which the ecological crisis has brought philosophical thought has made them look like a way out. The pending planetary catastrophe has raised anew the questions of whether there are alternatives to the kinds of economy, politics, science, and religion by which it was engendered as well as to the entire modern variation of thought that they constitute and cohere in, and thus too of whether it is in and through such nonmodern variations of thought that one might discover those specific alternatives. The anthropologists and philosophers contending with the crucial role played in the crisis by modern technics are asking whether and in what way other variations of thought might contain other technics – not just other arts of doing things but altogether distinct approaches to all arts – that render humans cohesive among themselves and with other beings, rather than working at all of their expense. Or, of how there might be, to use Yuk Hui's neologism, "cosmotechnics."

By showing in *The Question Concerning Technology in China* that the understandings of technics of Confucian, Daoist and other currents of the Chinese tradition are indispensable to that inquiry into technical variation, Hui has set a direction for it that leads right to the strange arts that anthropologists have for the last century called techniques of the body. Chinese thought from as early as Confucius and Zhuangzi to as late as the eighteenth century conceives of technics, on Hui's interpretation, as in essence naturally and morally uniting humans with the immanent process of the Dao, and much Daoist thought verges on making the paradigm and basis of that technics the arts of movement, breathing, and vitalization that go today by names such as *wai gong*, *nei dan*, and, generically, *qi gong*, and that blend into traditional Chinese medicine. Thus Hui's demand that philosophy ponder a variety of technics that opens a thoroughly different relation to the world than do industrial and digital techniques also stimulates us to consider that they could be put on equal footing with – and even subordinated to – arts so simple as to never reach past the body.

Now that possibility also confronts us with the further, outlandish prospect that the body, too, might be several, and that some of its variations may disturb our certainty about just *what* it is that Chinese and other technics work on and from in the first place. This is because the body of the Chinese corporeal arts presents the quandary of something almost impossible in its ontological incongruity, a thing that is, on the one hand, often indistinguishable from much else in a cosmos that is in turn composed of elements and relations – the subtle "energy" of *qi* especially – that are often not quite material nor at all real; while at the same time being, on the other, the very "object" that those arts presume and affect in order to produce their cures and changes. As nearly all science and most philosophy is unlikely to be willing or able to recognize that this unreal body might have to *exist* if it is indeed the actual target and the source of efficacy of those arts, it falls to the kind of thinking that has sought to displace the most unquestionable of modern concepts – the strain of anthropology that has dared to turn its relativism on even physical nature – to contemplate how this could be. But to do that, this anthropological relativism must be directed not only at some of the modern *concepts* by which the Chinese body ordinarily would be understood (the approach until now) but at the very *reality* to which they correspond and by which it is relegated to nonexistence. That double approach – an anthropology of variations of both thought and thing (or "heterology" meets "polyrealism") – may reveal the existence of that body and that too of its technics.

Or, to put all of this in terms that state the stakes and the itinerary of this essay, to ask the question, "What can a body do?" in light of Chinese cosmotechnics requires addressing those of "Which body?," "By what art?," "In which 'reality'?," and, in sum, "How might a cosmic art do?" I do this by arguing, in a series of discussions that mix commentary, description, and speculation, (1–2) that a pair of concepts that would have been thought to provide answers, *habitus* and Mauss's notion of "techniques of the body" respectively obscure the cosmic character of the body of the Chinese corporeal arts and the possibility of *cosmic* techniques; (3) that ontological anthropology renders both such cosmic techniques and cosmic technics conceivable by making unreal entities and processes integral to their definition; (4) that the unreal thing called *qi* must be conceded some kind of existence if the efficacy of the Chinese corporeal–cosmic arts is to be explained; and (5) that those arts may provide the basis for or at least access to another, nonmodern technics by allowing their practitioners to yield to and thereby work from the Dao and with things and processes.

## bodies more than bodies

That there are other conceptions of the body than those of European modernity is not controversial, but the idea that they are so divergent as to render nature, science, and technics relative is. Since Nietzsche and Freud and de Beauvoir and Fanon forced philosophy to acknowledge that the human body is among its sources and chief problems, it has become common to speak in the singular of the various bodies that were marked in the long colonial-modern era – by their gender or lack of it, or the race projected onto them, or their laboring character – as the sheer other of rational consciousness. It does not matter if those bodies are conceived of as primarily material, or discursive, or as historical chiasms of the two; they are thought to be instances of a finally unvarying human body.

That this body of critical thought too might be a variation among and *of* still other bodies, the counterpart to many other "bodies" and thinkable only from the perspective of each of them – that is entirely another thing! The mere thought of it seems eccentric, unneeded; it likely only dawns on people, bookish or not – intersexed or disabled, from cultures deemed inconsequential, or who work through their racialization by divinizing their flesh – who have felt the misfit between even the critical body and some other, whether their own or that of someone else. Some authors and readers of anthropological thought might be counted among them (on this point, they sometimes are among the estranged), as that incompatibility is made obvious by texts from its corpus in which not even an uncannily familiar body – nor sometimes any body at all – can be found. There are bodies that host several persons, others that pair up to make a single person, and still more that are the sum of their metamorphoses across species and human kinds. In the early French ethnologist Maurice Leenhardt's celebrated *Do Kamo*, its author reported that his New Caledonian interlocutors said that theirs are composed more of the stuff of trees than properly human flesh (16–19). But to call these bodies is probably wrong. Leenhardt's question of whether French colonization brought to New Caledonia the mind elicited from one Bousoou this still-astonishing retort: "Mind? Bah! You didn't bring us the mind. We already knew the mind existed [...] What you've brought us is the body" (Clifford 172). It may look to be much closer to some of us for appearing to correspond to the human individual, but the body of the Chinese corporeal arts is nevertheless as dispersed into the cosmos and at points indistinguishable from it. By the difficult cosmology that guides workaday traditional Chinese medicine as much as the esoteric Daoist arts, the body is never quite the physiology recognized by biomedicine but rather a distinct, often unreal anatomy of acupuncture meridians and points, tangible organs and intangible centers of vitality, and qualities of *yinyang* and fire, wood, metal, earth, and water that connect them in relationships of correspondence among themselves as well as with the rest of the "ten thousand beings."

In spite of their awareness of the equivocations that this body and others introduce into the modern concept of the body, anthropologists have preferred to maintain its univocal meaning by conceiving of "the" thing itself as a unity of sociophysical aptitudes reciprocally engendered by and engendering of its historical conditions. The inventors of that version of the modern body gave it the ancient name *habitus* in order to convey, as one of them – Pierre Bourdieu – put it, that the each and every body is a locus of "durable, transposable dispositions" that function "as principles of the generation and structuring of [the] practices and representations" that constitute society; a carrier of prereflexive cognitive-motor schemas by which so-called agents are able to act in and grasp their world because its very structures are formed by the slightest of its gestures (Bourdieu 72). That conception may be suited to grasping it as the site of the interplay of objective and subjective forces in social reproduction, but it rests on the colonial-modern supposition of a universe in which individual beings are discrete to themselves and discontinuous with each other and where human beings grasp that by naturally recognizing the gestalt image of the unity of their physical body.

Neither that world nor its body is arranged like those of the Chinese corporeal arts. No doubt their practitioners have a fine (often superior) sense of the body as a physiological and kinesthetic whole that acts and is acted on as a unity, but the commensurability between their broader understanding of it and that of social theory ends where the multiplicity of cosmic, often quasi-physical elements that they think composes it begins. In traditional Chinese medicine, the *mo*, or "arteries," are not a unified system that radiates from the heart but a network of independent pathways that are each specific to one of the *zang*, the "five organs" of the heart, spleen, lungs, kidneys, and liver; the *jing luo*, or system of "meridians," along which *qi* travels is even more decentralized, as none of its six primary pairs governs any of the others; and groups of *mo*, meridians, and organs share an essence with their counterparts in the seasons, land, plants, and so on. *Qi gong* practices complicate things by emphasizing three even more unreal entities from that system – *jing*, the physical essence of vitality; *qi*, its subtle, quasi-gaseous state; and the *shen*, or, roughly, "spirit" or "psyche" – while adding others, including additional meridians and a repository of *qi* beneath the navel called the lower *dan'tien*, and amplifying their continuity with the cosmos until they seem to be out in it.

It would be not only unimaginative but a bit crude to join past anthropologists in assuming that a person working (especially as a matter of unconscious habit) with such a plural and diffuse picture of the body would act primarily in ways that reflect the image of it as an isolated physiological unit moved only by dispositions it acquires from operating in a purely sociohistorical world. As practical as traditional Chinese medicine may be, excellence in it is based on not only the tacit motor and perceptual skills needed to treat an organ-*mo*-meridian block with herbs, acupuncture, and moxibustion but also a capacity to sense the resonances between it and the aspects of the cosmos to which it corresponds. In *qi gong* techniques that focus on the "spiritual," alchemical transformation of the self, the physiological dimensions of corporeality are often more radically deemphasized in favor of (apparent) intangibles such as *qi* and their circulation through the meridians and into the *dan'tien* in order to bring the system into balance. In view of both things, it is unlikely that the gestalt of the lived physical body per se is the chief point of reference in these corporeal arts or that it provides much of a description of the cosmic thing that instead might be.

## *les techniques du corps*, or: cosmic techniques

The modern belief in the unicity of all bodies also crucially determined the canonically anthropological notions that such arts are techniques of the body – arts of effectuating things by means of bodies that also inculcate sociality in the individual – and that, by Marcel Mauss's formula, "man's first and most natural technical object, and at the same time technical means, is his body" (83). As self-evident as those ideas have seemed to so many of the sociologist's heirs, they distort techniques that may be far more of the cosmos than the body.

"It is possible," Mauss states at the outset of the eponymous classic of social theory in which he invents this concept, "to produce a theory of *the* technique of the body in the singular on the basis of a study, an exposition, a description pure and simple of techniques of the body in the plural" (77–78). Which techniques and which bodies provide the basis for his definition of "the" technique of the body as such is evident in Durkheim's nephew's account of the staccato, half-conscious process of inductive reasoning by which it dawned on him that the body has techniques. He realized that the differences in how generations of the French swim, European militaries march, and urban nations walk indicate the social origin of bodily habits, and then that every such "acquired ability" or "faculty," which Aristotle named *hexis*, should be called the roughly equivalent Latin *habitus* rather than the too-psychological French *habitude* so as to reflect its entrained and social character; that each *habitus* is engendered through a threefold process of education by which the individual is psychically impelled to imitate the acts of prestigious individuals, which in turn activates and socially forms biological potentials of the body; that ethnological reports of indigenous Australian hunting practices that are as much ritualistic as means-ends in character attest to the interpenetration of magic, religion, and technics in that and other such instances of *habitus*; and, finally, that such practices are techniques despite not involving instruments, but that this has been unapparent due to the confusion in the philosophical tradition of technics as such with prosthetic techniques:

> I made [...] the fundamental mistake of thinking that there is technique only when there is an instrument. I had to go back to ancient notions, to the Platonic position on technique, for Plato spoke of a technique of music and in particular of a technique of

> dance, and extend these notions. I call technique an action which is *effective* [*efficace*] and *traditional* (and you will see that it is no different [*sic*] from a magical, religious, or symbolic action). It has to be effective and traditional [as] there is no technique and no transmission in the absence of tradition [...] But what is the difference between the effective traditional action of religion, the symbolic or juridical effective traditional action, the action of life in common, moral actions on the one hand and the traditional actions of technique on the other? It is that the latter are felt by the author as actions of a mechanical, physical, or psychochemical order and that they are pursued with that aim in view. (Mauss 82–83)

That is, it was the bodily habits of the European peoples most familiar to Mauss that led him to realize that they are social, that they ought to be conceived in quasi-Aristotelean rather than subjectivist terms, that they are psychobiosocial processes of education, and that they are technical despite not being prosthetic in character. While it was, on the other hand, his additional consideration of Aboriginal hunting practices that leads him to decide that the techniques may also have supplemental magical, religious, and symbolic dimensions. The practices of many other peoples – of South Asia, the Andes, Central Africa, Tierra del Fuego, the Maghreb, the Southwest of the United States, Polynesia, and China, all *c.*1934 – that subsequently figure in his inventory of techniques of the body play little role in his definition of the techniques as at once effective and traditional.

The surprisingly European character of the bodies that Mauss finds obtruding into the concept of technics allows him not only to equate the body as such with the individual physiological body but also to reduce many of the arts that he avows have a "magical" dimension to techniques of producing with that body mechanical effects. For instance, commenting on the Hopi and Pueblo arts of running in his discussion of techniques of sleep, rest, movement, and reproduction might have easily prompted someone of Mauss's erudition to consider that those very peoples and most of their neighbors indeed had sleeping arts – arts of dreaming – but that these brought forth effects that were less physiological than psychic by way of incorporeal, "inefficient" causal processes. The very indigenous Australian hunting practices by which so much of his analysis is forwarded, moreover, exceeds Mauss's framework even further by being undertaken less by a human being than an agency that is a massive aspect of the entire cosmos. The sociologist reports that in a particular kind of possum hunt, the hunter not only chants a magical formula while pursuing his quarry but holds in his mouth a quartz crystal – an instrument, or perhaps instead a sentient body – while doing so. There is much in anthropological literature to indicate that this mineral is likely to have been conceived of by the hunter as amplifying his power by virtue of its connection to his Dreaming-being, or "totem," and the group of beings that share together its essence, which may very well mean that he also would have concurred with indigenous reckonings that the beings in such "classes" can affect each other in an extramechanical fashion. Anthropologists like Barbara Glowczewski suggest the implications of this for the status of the hunt as a technique by showing that Aboriginal peoples still often conceive of themselves almost as *instruments* of the Dreaming-beings' way of reproducing their creation through humans, while indigenous thinkers such as Big Bill Neidjie and Aileen Moreton-Robinson speak of the Dreaming as a pervasive, quasi-causal force that affects everything on native land and empowers or overrides that of its inhabitants.[2] To follow Mauss by supposing that actions of "the body" undertaken according to that understanding are corporeal and technical in any familiar sense is to flatten it into a modern cosmology in which its own notions of process, individual, and agency are legible only as inaccurate figurations of real causal relations between real beings.

Of course Mauss was not uninterested in such vicissitudes of causality, as he invents with Hubert a distinctly social notion of it in *A General Theory of Magic* in part to resolve the quandary of the efficacy sometimes evident in sorcery. But just as with the great Maussian concepts of the gift and the person, the radical possibilities of that of body techniques are kept in a state of penumbral latency by the sociologist's judgment that they are universal to humanity. That view has proved acceptable enough to anthropologists that they have inherited it without question across several otherwise quarreling generations, and it has led even some of the most critical among them to assume that many of the religious, performative, and disciplinary practices that they define as ethical regimes are foremost techniques of the body. This has made it seem that so-called ethics of the self are undertaken primarily if not exclusively through work on the body as it is conceived by modern thought – through exercises that are in the first instance physiological and only afterward psychic or spiritual – and that the possibility that they might also or more often take place through whole swathes of the cosmos is the sort of extravagance fabricated by overly imaginative speculative minds. That it would be just mad to think that some techniques of the body are far more of the cosmos, that cosmotechnics are better to think ...

## other ontological distributions, other technics

The interest of the failures of anthropologists to register the disparities between other concepts of the body and technics and their own – the reason that anthropological thought has to be engaged here rather than primarily philosophy, critical theory, history, or the neural sciences, or just other variations of thought "by themselves" – is the transient but radical improvements on themselves to which these errors gave rise. Among them is the turn of some recent anthropologists toward conceiving of the body and much else that has been thought to be invariably natural and/or cultural as instead ontologically varying. In contrast with the sure directions pursued by many of their peers, their detour leads to a way to conceptualize techniques – and whole technics – conducted from the cosmos.

This turn to ontology is arguably initiated by Claude Lévi-Strauss in a line of thought that runs from *Race et histoire* through *La pensée sauvage* and the *Mythologiques*, and perhaps foremost in the possibly apocryphal anecdote about early colonialism that he recounts in the first of those texts. At the same moment when early Spanish colonists were torturing Taino natives of the Antilles to determine if the theological opinion that they lacked a fully human soul was correct, the Taino were conducting a parallel experiment on their own, Spanish captives to determine if they possessed the same kind of body as themselves (Lévi-Strauss 329). Lévi-Strauss hears in the story a parable saying that the denial of humanity to the other is the trait most common to humanity, but he also can rightly be said to have unwittingly registered that its message is instead that nature is among its most uncommon concepts. As Eduardo Viveiros de Castro famously argued, the Antilleans' reason for pursuing their strange investigation is that they likely shared with certain peoples of tropical South America, the American and Siberian Arctics, and Central Asia the notion that most if not all beings (human by modern standards or not) are entirely human and cultural by virtue of having the same kind of soul and are distinguished into kinds not (as in the modern multicultural imaginary) by any such spiritual trait but by their specific kinds of bodies: by his reckoning, a people holding that supposition about the cosmos will regard other human collectives as being no less different to itself than animal species and contend with that alterity by attempting to make sense of their specific kinds of bodies and the unique perspectives that these grant them. The significance of this for the Brazilian anthropologist is that it thoroughly upends the assumption that culture and nature are recognized in all human thought and exposes the local character of that modern distinction. For even as this concept of the body as speciating differentiator rather than differentiated substrate was detectable by virtue of Lévi-Strauss's principle that kinship systems, myths, and other anthropological phenomena are identifiable only through comparison, it deprives that method of its ground by revealing that the modern distribution of beings into human and nonhuman that leads us to call it a cultural representation is profoundly alien to it. In conceiving physicality not as the homogenous given on which diverse conventional representations are instituted but the means by which perspectival heterogeneity is introduced into a uniform human universe, these "multinaturalist" cosmologies prove to distribute beings in a way – to work with an ontological distribution – that is distinct from (even the inverse of) that of modernity.

The relativization of the body and nature that Lévi-Strauss thus sets in motion also reaches to and affects technics. In reckoning in *Beyond Nature and Culture* with the implications of both structuralism and Amazonian cosmology for ontology as such, Philippe Descola argues that the character of any technics is found in a broader ontological distribution. In the course of elaborating there a theory of how it is that ontological distributions come to be altered, the Chaire d'anthropologie de la nature states that Amazonian peoples have often shown indifference to techniques incompatible with their own such distribution. Although groups like the Achuar made use of metal tools, firearms, motors, and machines for their forging and repair introduced to them by whites, they did not adopt apparently simpler and less exotic techniques for the domestication of animals as livestock. This is due, Descola contends, not to the absence of customs of taming – individual cubs of peccaries orphaned during hunts, for instance, are raised and even nursed in villages, where they roam freely – but to the normal status of animals as human collectives that can be hunted and eaten only on the condition of remaining aliens. Tamed animals are, by contrast, not consumed because they have been virtually adopted into the collective as kin, and the domestication of whole species likewise represents the prospect of reducing their alterity to the point that they could be consumed no more easily than fellow tribespeople. Such animist conceptions of the animal qua foreign and often enemy person only seem to dissolve when a whole species becomes domesticated for contingent reasons, as that single exception leads to others that weaken the overall sense of the parity of animals with humans. This and other cases of indifference to obviously beneficial modern techniques lead Descola to conclude that "a technical relation is remarkably stable over time, in contrast to the instrumental means that it employs [...]"

> The range of possible relations with oneself or some other living or inert being is far more limited than the whole collection of objects that those relations may engender and than the gamut of usable methods to achieve one's aim. Technical evolution could therefore be envisaged, not as a gradual complexification of tools and processes of transformation, but rather as a limited and more or less cumulative series of objectivations of new relations. The transition from cart to automobile or from the belted loom to the Jacquard loom was in no way inevitable or predictable, but at least it seems compatible with the features of such operations and the nature of the results inherent to those categories of artifacts. However, that is by no means the case with objectivation of a new technical relationship, such as the domestication of plants and animals [...], which constitute unprecedented revolutions in the apprehension and treatment of the frontiers between oneself and others. (386–87)

Should every technical relation indeed express a specific ontological distribution, then the broader technics that comprise those relations can and must be defined to include the very sorts of psychic and cosmic arts that are ordinarily treated as merely imaginary, magical supplements to its corporeal and prosthetic core. To continue with hunting in Amazonia but now as described by Davi Kopenawa in *The Falling Sky*, Yanomami peoples rarely get underway with it unless a shaman like himself has first secured rights to certain kinds of game by calling forth and making alliances with their corresponding *xapiri*, the "spirits" that are nearly the most fundamental (person-) forces in the cosmos. Yet as much as a shaman relies on crafting and ingesting psychedelic substances like *yakona* to enter those negotiations, the extracorporeal and extrainstrumental dimension of hunting cannot be limited to that pharmacological practice: skill at visions and dreaming is at the heart of the métier of the shaman, while his ability to interact with the *xapiri* comes from having already transmogrified his body into the dwelling of a vast number of those agencies and thus a swathe of the cosmos. The specific technique of which hunting is a part therefore should neither be confined to tracking and weapon-use nor distinguished from the dreaming arts of the shaman and the hunters following his guidance, while their bodies and those that they target ought not to be dissociated from the psychic dimension of the cosmos that "causally" determines them. As awkward as it may be to adopt Kopenawa's point of view, doing so exposes the distinctly animist character of the technics of which "hunting" is an aspect, that it intervenes at the level of the human souls that bodies conceal and where beings are deemed to be most susceptible to affection.

Yet to acknowledge that an extramodern technics may be configured to and made from beings and modalities of being that have no place in the physical universe requires taking the further step of clearing another, pluriversal kind of space in which they can find room. Both Viveiros de Castro and Descola have done that by revisiting the classic distinction between primary and secondary qualities on which the sciences still effectively draw, although their interest in it is just the opposite of the lights of realist philosophy who parallel them in this: not to confirm with it some version of modern thought but to show that the "qualities" at the bottom of everything – or at its peak, or in its wellsprings, or where it is most potent – are often entirely other for other variations of thought and must be at least interpretively affirmed if one is to follow how they conceive of doing things with and from the cosmos.

## an art of unreal existents

That metaphysical decision – those anthropologists' refusal of one of the conceptual distinctions that holds modern thought together, and acceptance of the responsibility to affirm and construct others – is almost inconceivable to philosophers today. Their concern is to define the basic properties of the universe or the appropriate use of a reason unable to touch them as they are, not to ask what there might be and how it is once other variations of thought are reckoned with. Most of the supposedly self-evident reasons that philosophers have for not thinking the outside of their tradition (that philosophy is foremost of logic and science, that it is as such critical, that relativism is its foremost enemy, that its mission is to adjust us to a unitary and indifferent cosmos) therefore are of little relevance here, but their likely objection that the *existence* of other bodies, cosmoses, and technics does not follow from concepts of them must be acknowledged.

The ontological anthropologists have, unsurprisingly, not tried to say otherwise and have accordingly limited their inventories of the other beings that are perceptible and conceivable in nonmodern thought to those, like animal subjectivities or the resemblances at the basis of analogies, that are insignificant but not unreal to the sciences. Their prudence should not lead one to believe, however, that there is a line where questioning becomes untenable and comes to an end – a bottomless edge of thought guarded by unquestionable things – and that it lies where most philosophers say it does, as it quickly vanishes if the sciences are not passively conceded the power to make final determinations about what there is. As long as it is accepted that there are mutually irreducible variations of thought and that they must be understood on their own terms, then there are at least questions about false, preposterous, imaginary, and other unreal things to be asked, and they cannot be ruled out in advance.

In respect to bodies that are more than bodies, there is a simple question, about their existence – about whether some of them have tangibility and presence – that shows that such inquiries lead somewhere. Almost everything in even the most open and creative intellectual disposition protests that lending thought to such a matter is some combination of inadvisable, misguided, stupid, and plain mad, but that seems correct only because we have become too accustomed to reality and the respectability that recognizing it ensures. That concept (when it is even articulated as one) deserves far less regard, as it is most often little more than an ultimate signified that is believed to always coincide with the totality of probable and verified referents of language. There is much else that exists that exceeds the real, and not all of it is intangible.

The "body" of the Chinese corporeal arts offers disconcerting evidence for such unreal existents, as some of its most apparently symbolic or erroneous aspects evince an improbable but insistent tangibility and affectability. As Shigehisa Kuriyama shows in *The Expressiveness of the Body and the Divergence of Greek and Chinese Medicine*, the *mo*, or (here) "pulse," detected by traditional Chinese medicine's art of palpitation differs from the pulse of biomedicine in being found at three locations on the wrist as well as places on the body lacking blood vessels and through binary qualities, like floating or sunken and slippery or rough, that are at best poor metaphors by which to measure heart rate and blood pressure (17–108). Yet that neither makes the *mo* nonexistent nor undermines its diagnostic value. Although colonial English doctors found the terminology of Chinese medicine to be metaphoric to the point of unintelligibility, they acknowledged that their Chinese counterparts were achieving valid perceptions in pulse-taking because of their evident ability to make with it accurate and sometimes better diagnoses than themselves. A sunken pulse in the third position of the right wrist, for example, might index compromised *yang qi* in the kidneys and thus "in fact" dysfunction in them.

Yet even if the *mo* is acknowledged to be a concept and a percept distinct from the pulses of biomedicine and its early modern antecedents, its referent and object can be taken to coincide with throbbing blood at the radial artery, and it thus does not quite constitute a case of an unreal but possible existent. In spite of confining himself to a history of divergent kinds of perception rather than of things, Kuriyama points out that blood, by contrast, is not in essence conceived of as distinct from *qi* throughout the various periods of Chinese medicine but rather is the formed, *yang* expression of a general vitality of which *qi* is the formless, *yin* other side, and that a more extensive idea of physicality as the outer face of the vital also holds for much of the rest of the body: neither the five *zang* of the heart, spleen, lung, kidneys, and liver nor the minor *fu* such as the stomach, large and small intestines, and "triple burner" are organs in the sense of discrete anatomical entities with specific functions, as they are rather repositories of *qi* that convert it into powers unique to themselves; while the meridians and points are not nerves and their endings but the routes along which *qi* travels in partial parallel with blood and the places where it pools (265–66). The discrepancy between these two groups of corporeal notions is so thorough that it will seem that the one of Chinese medicine could symbolize the body that biomedicine's accurately references. Yet *qi* is understood by practitioners of the corporeal arts to be something replete with positivity, a felt existent that is certainly not as evident as are visible and audible things but that is reliably tangible to those who have learned to perceive it. Farqhaur and Zhang share that that "the acupuncturist only knows that she has found the right acupoint when she can palpably 'get *qi*'" and that "the *qi gong* adept learns to feel *qi* as a specific resistance in the air or as a wind traveling between bodies"; Chen, that patients sick from incorrect *qi gong* practice sense *qi* moving violently through them; and Kaptchuk, that doctors of Chinese medicine easily distinguish between *qi* that is flowing or stagnant, expanded or flat.[3]

Such perceptions pose significant trouble for reality, as they can scarcely be excised from an account of the formidable abilities in diagnosis and treatment of practitioners of Chinese medicine. Their demonstrated successes – with not only anesthesia and blood pressure but also blood, liver, and other major diseases – may be attributable to herbal pharmacotherapy and the capacity of acupuncture to stimulate the nervous system and the brain, but even the most educated doctors of Chinese medicine often presume that affecting *qi* is just as crucial. By their own accounts, there is something there, a pattern of sensations of an apparent entity or force that may be uncertain in this or that instance but that reliably presents itself, behaves in an anticipatable manner, and that is susceptible to intervention. That view is likely as much if not more than premodern cosmology the reason for the attempts of scientists in China in the 1970s and 1980s to verify by measurement the hypothesis that *qi* is bioelectricity or some other known physical phenomenon. That research may have yielded few convincing results and no unanimity about its conclusions, but the existence of an almost spectral, extraphysical aspect of the body was never in doubt for those practitioners, and that supposition is awkwardly indigestible by not only scientistic but even the most critical thought.

The prospect that things are not limited to being physical-natural or linguistic-cultural or some amalgamation of both, however, changes *qi* from an obtrusive but dead-end mystery to a live problem. Even if the idea that other variations of thought are irreducible to that of modernity is no more than a speculation based on some metaphysical reasoning, it holds strong consequences for the natural and human sciences, the philosophies that issue rulings on them, and the modern ontological distribution that still holds sway over them both. Should other variations of thought be approached on their own terms, the reports of their learned adherents of perceptions of and interactions with beings that are flat out impossible to the sciences ought not to be excluded from intellectual consideration. It would instead be better to grant enough credence to other formats and methods of perception to accept that they may reach perceptums not falling within the spectrum of perceptibility accepted by educated common sense. As inherently unfruitful as such a stance toward the unreal might seem, holding it reveals that the *un*scientific criteria for good perception of those other methods nevertheless may be of such refined specificity that it makes little sense to assume outright that their "objects" are just the symbolic décor of tradition.

In traditional Chinese medicine, the criteria for a worthwhile, actionable perception could be said to be to this day not locality, genericity, and acuity but globality, singularity, and perspicuity. Kaptchuk says of Chinese medical examination, for instance, that "a truly adept physician can diagnose from almost any single piece in the human configuration," for in "the finest shadings of a particular emotional ambience, or tongue, or manner of waking, or pulse, the master physician can discern a pattern" that holds for the rest of the body (285). Where *qi* in the raw is concerned, perception is, on many accounts, additionally vague and diffuse rather than clear and focused. In acupuncture, "getting the *qi*" is a matter of feeling slight tension at the needle and sometimes of catching the feeling of something like a current or fluid passing through the acupoint. In *qi gong*, such perceptions become paramount, but are achieved not by searching for them but by resting one's attention on one's physical movements, breath, or *dan'tien* and then allowing for a passive awareness of the circulation of *qi* as well as sensations of heat in the places where it is thought to gather. As not everyone perceives *qi* initially or even for some time when practicing, some teachers instruct that it is by entering a meditative, detached state that accuracy can be forgotten in favor of such oblique perception.

As long as such perceptions are accepted, then *qi* can and ought to be granted existence in spite of not being real. That is, an entity that is so subtle as to be "dubious" but not dubitable (i.e., unlikely but almost unamenable to hypothesis) – a presumed thing, or force, or entirely other sort of *x* that is usually imperceptible to the senses and instruments and likely uncapturable under experimental conditions – is nevertheless reliably there to learned people who can cure by affecting it, and it would be wise to concede rather than to deny their claims to its existence unless one can guarantee that sense perceptions are always of objects of nature.

Such a choice of course deprives one of a sure cosmological orientation as well as the mystique of disenchanted rationality that is derived from regulating statements about nature and history. Yet that loss in certainty about the space and purpose of thinking is a massive gain for it in possibility, as the long shadow that reality has cast over being in our era can be seen through once its source has been distinguished from other, more captivating features of the surrounding expanse.

## "technics" from dao

It could be easy to believe that whatever beings might be brought back into view this way are objects about which knowledge can be inferred through familiar kinds of reasoning; here, that whatever sort of existence *qi* could be said to have could be extended to the meridians, the *jing*, the *shen*, and the entire body of Chinese medicine of which they are aspects. Yet according to a line of philological argument of François Jullien's that is as relativistic as the most radical anthropology, *qi* makes "body" altogether fail as a translation for that group of concepts, and the thing that it evokes instead has the sort of thoroughly cosmic sense that by now is to be expected. By following the sinologist's elucidation of the ontological character of that nonbody, those too of the arts and the technics peculiar to them can come to light.

In one of the phases of his virtually endless displacement of Occidental thought by way of that of China, Jullien shows that *qi* is conceived of in a text at the foundation of Daoism as not one being among others but the most subtle, virtual state of all beings, that in and by which they are nearly coextensive with the Dao. In the long commentary on Zhuangzi that is his *Vital Nourishment*, Jullien introduces this idea by remarking that the colloquial Mandarin expression *yang shen*, "to feed one's life," is best understood as concerning nourishment in neither the literal nor figurative senses of the feeding of body or soul but rather through the fourth-century BC sage's notion of it: the cultivation of one's capacities by breathing *qi* and thereby becoming indistinguishable from it and the Dao. In contrast with Aristotle's precept in *The Nichomachean Ethics* that considerations of nutrition and perception must be excluded to obtain theoretical knowledge of the good, Zhuangzi states that intellection must be left behind in favor of a return to "the source of 'physical' being" if one's life is to be cultivated; to the *du*, the "principal artery" by which *qi* travels the length of the spine and that *qi gong* practices open with breath due to the meridian's capacity to affect the rest (Jullien, *Vital Nourishment* 16). As much as this ancient saying may seem to be the injunction to attend foremost to one's body that it is often understood to be, Jullien shows that there is no cognate for Occidental notions of the body in Zhuangzi. *Shen* and *ti* respectively mean something like "animating self" or "person" and "constitutive being" rather than body, while the apparently closer match of *xing*, literally "shape," is, according to the sinologist, best rendered as "particular actualized form," as it means human or animal physiology only sometimes and more often just the shape peculiar to a thing of any kind (66–69). As such forms result for Zhuangzi not from a soul impressing itself onto matter but from the undivided cycle by which the undifferentiated Dao constantly changes into *qi* and then individuated concretions of it before the process reverses, they are only partly and superficially distinct from the rest of the cosmos. Far from being the physical aspect of a person, the "body" is a transient instance of the Dao.

This indistinguishableness from *qi* and Dao of the *xing* applies too to the act by which the corporeal arts nourish it. This emerges, on Jullien's reading, upon examination of the way in which Zhuangzi speaks of breathing. He does this through an allusive style of discourse ("detour") that the sinologist also ascribes to Confucius and Lao Tzu and that consists in evoking but never providing the meaning of a concept by continually varying it. When Lao Tzu successively characterizes the Dao as "great," "receding," "distant," and "returning," for instance, each term's signification defers or diverges from that of the preceding and thereby conveys that the Dao is never exhausted by the actualizations it becomes. Zhuangzi employs that method to convey that *qi* is too virtual and limitless to be defined or located, and that the breathing by which one gathers it is likewise a process that cannot be said or found to occur in this or that body and place and that is thus indistinct from the cycle of gathering and dispersing that is the Dao. In other words, Zhuangzi indicates that the practice at the heart of *qi gong* takes place at once with and *as* the cosmos.

The importance of this effectively ontological characterization of those arts is that it pertains not only to them but also to how they belong if not provide access to an entire technics. Apropos the celebrated tale of Butcher Ding from the ninth chapter of Zhuangzi, Jullien argues that the butcher's explanation to Prince Wenhui that his extraordinary abilities come from the Dao rather than his own skill must be understood in light of the role he accords in that to living off *qi*. Ding states that where the good and the mediocre butcher must respectively change blades every year and month on account of the one's regularly cutting flesh and the other bone, he has employed the very same tool for nineteen years because

> My contact with it is [...] spiritual, and I no longer see it wholly with my eyes. When the knowledge of the senses ends, my spiritual faculty seeks to go further by attending to the natural ['heavenly'] structure of the animal. So I attack the big spaces and guide my blade through the broad passages by hewing along the internal shape. I therefore never touch the veins, arteries, muscles, or nerves, much less the big bones. (Jullien, *Vital Nourishment* 88)

Ding could be taken as saying that it is his awakened mind or soul that allows him to effortlessly, perfectly slice through the densely variegated carcass by yielding to it rather than providing it with a force to resist. Yet Jullien stresses two features of the passage that indicate that the butcher derives this ability from nourishing himself on *qi* and thereby connecting with the Dao: the butcher calls that "faculty" not mind or spirit (*xin*) but simple, unimpeded perception; and his description of it as connecting to an "internal" structure or shape suggests precisely the otherwise imperceptible cohesion of things that should be revealed by feeding off breath and *qi*. Prince Wenhui's astonished response of "[Now] I understand what it means to feed one's life," confirms, adds the sinologist, that the story concerns this sort of cosmic nourishment as much as the efficacy that issues from it. Its lesson is that it is by learning to let *qi* gather and disperse through breath that one comes to discern the interconnection of things and "the way" through them.

That the cosmic arts bring the individual to stop trying to transcend things and just yield to them finally affords an understanding of how they may provide the basis for a distinct technics. Jullien recognizes that their technicity – if that is the right word – and that of any art undertaken by way of them does not lie in their own power and manner of causing an effect, as whatever one intends to accomplish with them primarily will be because they allow processes to play out with minimal or no outside intervention. To illustrate, Zhuangzi elsewhere recounts an episode in which Confucius spots a man amidst river rapids lethal even to giant tortoises and crocodiles. Upon being asked later by the sage by what *dao*, or procedure, he is able to survive, the swimmer replies that he has none and instead yields to the *dao* of the water, letting it carry him where it will without resistance. Like Ding's, his skill is so subordinate to the things it engages that he need not actively direct it, and they accordingly bring about the desired outcome.

"Propensity" is the name that Jullien gives to that processual causality in *A Treatise on Efficacy*, the work that he devotes to this nonart of practicing any art. On the view of so many of the Chinese classics, every situation in warfare, politics, commerce, and so on is a complex of forces that has a propensity to bring about certain outcomes rather than others, and it is best to adapt to and work with it to achieve what one intends rather than to intervene upon it from even a moderately external position. That approach succeeds, Jullien observes, not just by refraining from manifest action and thereby not soliciting the resistance and reticence that lessen its effects but more crucially by being conducted from the level of the situation that is the least actualized; where the effects it is already beginning to engender are so nonspecific as to be nearly unpreventable (like a cornered, fearful army, which should not be provoked into valor by attacking but allowed time to lose morale so it fights poorly or just surrenders). Operating from that maximally immanent level of things is thus a matter of being adjusted not only to causes other than oneself but to *effects* in their most potent – because least actualized – state. The sage, or master of that cosmic nonart, operates in rapport with the Dao by connecting with some of its most virtual, latent, and indefinite effects.

## enduring cosmic stuff

In *The Question Concerning Technology in China*, Hui parallels anthropological thought in his manner of contending that this understanding of technics as a conduct of techniques by way of the Dao offers an alternative to that of most philosophy. As apparently irrelevant to a critique of modern technics as conceptions of technical objects from the Chinese tradition might seem, they reveal their intellectual potency when translated not into but rather against the modern concepts by which they will inevitably be evaluated. In calling Confucius's guidelines for reinforcing morality through the ritual use of objects a "technics," Hui gives the word the strange sense of an instrumental practice that by working from the Dao joins to it "humanity," which in turn becomes able to discern the cohesions and resonances between things. By producing an equivocation in the concept of technics in the way that anthropologists have with nature and culture, Hui exposes how technics has been otherwise and still can be.

In spite of his reliance on this approach to translation, Hui expresses reservations about the continued use of it in anthropology absent a confrontation with the omnipresence of modern technics and its destructive leveling of human thought. For him, a chief rationale for inquiring into the cosmology of China's or any other variation of thought is to learn how it might enable even the most hypermodern of techniques to be practiced differently – to interpret its cosmos and couple it to them – and to counter the fantasy that they can be rejected for a sheer return to nature. Such a wish for the sleep of a tranquil existence has often followed the harsh awakenings of peoples to the catastrophic fracturing of their cosmologies under colonialism, and its fulfillment has been repeatedly sought in reactionary and fascist programs to overcome modernity by restoring those orders to their supposedly archaic perfection. To believe with them that some homeland or even the Earth itself can be inhabited without craft and invention is to disavow the gap that not only Chinese but other variations of thought acknowledge between Dao and "human," and seek, in Hui's thinking, to bridge through technics. According to him, the affirmation of other cosmoses in anthropology risks succumbing to that denial by attributing to them a coherence that they have often lost to modern technics as well as a placidity that would make them havens from it.

As much as that caution ought to be heeded, it may confuse the pervasiveness of modernity with its power over other variations of thought and their cosmoses. In conducting its inquiries from outside the spaces ordered by modern thought rather than from within them, anthropology has been able to register when the other variations have not been made commensurate enough with it to lose their consistency or to be absorbed by it; where they have variously persisted, reconstituted, and recovered. By continuing with and intensifying here that method of thinking from the exterior, it has become not only clear that Chinese concepts of body and technics remain vital alongside one of the worst metastasis of modernity but also conceivable that they may maintain connections with things that so much knowledge has assured us do not exist. As inexpedient as such experimentation nevertheless might still seem, it may in time confirm that the very stuff of some other cosmoses neither were nor ever could have been eliminated and can be found as nearby as one's "body," and that the outline of part of a solution to the crisis of modern technics may lie in engaging in corporeal–cosmic techniques and becoming attentive to their effects. That will not suddenly displace the universe to which we have grown accustomed with a cosmos permeated with resonance and connection, but it may put the reality that modern technics obsesses us with in its place and allow processes and arts deemed illusions to reemerge and alter us.

## disclosure statement

No potential conflict of interest was reported by the author.

## notes

1 As I transliterate 氣 as *qi* here, the reader is advised that the word does not mean 器, "utensils," "instruments," etc., Hui's subversive cognate for "technical objects" in the Chinese context.

2 The formulation that humans are almost instruments for totems is Descola's (292).

3 Farqhaur and Zhang 259–60; Chen 77–106; Kaptchuk 41–74.

## bibliography

Bourdieu, Pierre. *Outline of a Theory of Practice*. Cambridge: Cambridge UP, 1977. Print.

Chen, Nancy. *Breathing Spaces: Qigong, Psychiatry, and Healing in China*. New York: Columbia UP, 2003. Print.

Clifford, James. *Person and Myth: Maurice Leenhardt in the Melanesian World*. Berkeley: U of California P, 1982. Print.

Descola, Philippe. *Beyond Nature and Culture*. Chicago: U of Chicago P, 2013. Print.

Farqhaur, Judith, and Qicheng Zhang. *Ten Thousand Things: Nurturing Life in Contemporary Beijing*. New York: Zone, 2002. Print.

Glowczewski, Barbara. *Desert Dreamers*. Minneapolis: Univocal, 2016. Print.

Hui, Yuk. *The Question Concerning Technology in China*. Cambridge: Urbanomic, 2016. Print.

Jullien, François. *A Treatise on Efficacy*. Honolulu: U of Hawai'i P, 2004. Print.

Jullien, François. *Vital Nourishment: Departing from Happiness*. New York: Zone, 2007. Print.

Kaptchuk, Ted. *The Web That Has No Weaver: Understanding Chinese Medicine*. New York: McGraw-Hill, 2000. Print.

Kopenawa, Davi, and Bruce Albert. *The Falling Sky: Words of a Yanomami Shaman*. Cambridge, MA: Harvard UP, 2016. Print.

Kuriyama, Shigehisa. *The Expressiveness of the Body and the Divergence of Greek and Chinese Medicine*. New York: Zone, 1999. Print.

Leenhardt, Maurice. *Do Kamo: Person and Myth in the Melanesian World*. Chicago: U of Chicago P, 1979. Print.

Lévi-Strauss, Claude. *Structural Anthropology II*. Chicago: U of Chicago P, 1976. Print.

Mauss, Marcel. "Techniques of the Body." *Techniques, Technology, and Civilization*. New York: Berghahn, 2006. Print.

Moreton-Robinson, Aileen. *The White Possessive: Property, Power, and Indigenous Sovereignty*. Minneapolis: U of Minnesota P, 2015. Print.

Neidjie, Bill, and Stephen Davis. *Kakadu Man*. Queanbeyan: Mybrood, 1985. Print.

Viveiros de Castro, Eduardo. *Cannibal Metaphysics: For a Post-Structural Anthropology*. Minneapolis: Univocal, 2014. Print.

# Index